The

CW00689824

About the Book

This revised edition is a guide to the care of the most popular children's pets. It has full-length chapters on the dog, cat, rabbit, hamster, gerbil, pony, budgerigar, tortoise and goldfish, with subsidiary chapters on the guinea-pig, rat, mouse, donkey, canary and terrapins.

Each chapter is a comprehensive guide to the many varieties of these pets available, and to their breeding, health, feeding, accommodation and routine care. The author gives a wealth of background information about the nature and habits of these very different animals, not only to inspire a sympathetic interest in them, but to provide knowledge for keeping them satisfactorily in captivity. The text is copiously illustrated with black-and-white photographs and drawings. This convenient sized pocket book is packed with as much information as one normally finds in a much larger format.

About the Author

Tina Hearne is a Regional Organiser for the RSPCA. A teacher by profession, she has lectured extensively for the RSPCA in schools and colleges during the last 25 years or so, and produced film-strips and slide-sets for the Society on pet care and natural history subjects. She also imparts her wide and expert knowledge to adults and children alike in her many articles and books, notably the official RSPCA Pet Guides, of which she is the author.

As well as the paperback *New Observer's* guides, there are hardback *Observers* too, covering a wide range of topics.

NATURAL HISTORY Birds Birds' Eggs Wild Animals Farm Animals Sea Fishes Butterflies Larger Moths Caterpillars Sea and Seashore Cats Trees Grasses Cacti Gardens Roses House Plants Vegetables Geology Fossils

SPORT AND LEISURE Golf Tennis Sea Fishing Music Folk Song Jazz Big Bands Sewing Furniture Architecture Churches

COLLECTING Awards and Medals Glass Pottery and Porcelain Silver Victoriana Firearms Kitchen Antiques

TRANSPORT Small Craft Canals Vintage Cars Classic Cars Manned Spaceflight Unmanned Spaceflight

TRAVEL AND HISTORY London Devon and Cornwall Cotswolds World Atlas European Costume Ancient Britain Heraldry

The New Observer's Book of
Pets

Tina Hearne

Illustrated with 32 black and white photographs
and with numerous line drawings by
Christine Bousefield

Frederick Warne

FREDERICK WARNE
Penguin Books Ltd, Harmondsworth, Middlesex, England
Viking Penguin Inc., 40 West 23rd Street, New York, New York 10010, U.S.A.
Penguin Books Australia Ltd, Ringwood, Victoria, Australia
Penguin Books Canada Ltd, 2801 John Street, Markham, Ontario, Canada L3R 1B4
Penguin Books (NZ) Ltd, 182–190 Wairau Road, Auckland 10, New Zealand

First published 1978
Reprinted 1980
Revised edition 1986

Originally published as *The Observer's Book of Pets* in small hardback format

ISBN 0 7232 1665 7

Printed and bound in Great Britain by William Clowes Limited, Beccles and
London

CONTENTS

INTRODUCTION

The choice of animals in this book is based on those most commonly kept, but these are not necessarily recommended species. It will be obvious to the reader, for instance, that the keeping of such exotic species as Mediterranean tortoises should not be encouraged. Indeed even that doyen of the pet world, the dog, cannot be recommended unreservedly. The present sad decline in the dog's reputation is mainly due to the keeping of unsuitable breeds in urban conditions. The first responsibility of pet owners is to make a sensible choice of pet, bearing in mind the limitations of their own living conditions.

It is true that there are health risks in keeping animals. Cases of rabies, toxocara, and salmonella poisoning attributable to pets are given much prominence in the press, but providing sensible precautions are taken, these risks should be minimal. Probably those in most danger are very young children who have not yet learnt the basic rules of hygiene, and continually put their hands to their mouths. They should be encouraged to wash after handling the family pets, not to handle strange or sick animals, and never to neglect a scratch or bite.

There is also the risk of allergy. Quite frequently a child will display an allergic reaction to a particular animal, with symptoms of coughing, wheezing, running eyes, and skin irritation. There is no remedy, and the only solution is to remove the animal immediately the child's distress is noticed.

It is taken for granted that all pets must be kept in hygienic conditions in cages cleaned with hot water and household detergent. Iodophor disinfectants such as Savlon are recommended and are available from chemists' shops. Iodophors have the advantage of losing their scent when they lose efficiency. Risks of infection are, of course, minimized if animals are bought only from well-known breeders or reputable dealers. Under no circumstances should an animal be taken directly from the wild and kept as a pet.

All animals described in this book are protected by law, and it is an offence to ill-treat them. Equally it is an offence not to provide for their needs. The owner has the so-called 'duty of care'. Leaving one of these pets without adequate means of support, if only for a weekend, is construed in law as abandonment, and the owner is liable for prosecution. Owners are also reminded that they are liable for any damage their pets may cause. In particular, the owners of ponies, dogs, and any other animal that may run out on to a road and cause an accident should have an adequate third-party insurance.

The RSPCA's inexpensive pamphlets on pet care are listed in its educational resources catalogue. Please enquire from RSPCA, Causeway, Horsham, RH12 1HG.

Warning Electrical circuits in fish and terrapin tanks should be installed or checked by an electrician.

ACKNOWLEDGEMENTS

The author and publishers wish to thank the following for their kind permission to reproduce photographs: Sally Anne Thompson, Animal Photography Ltd for pages 12 (all), 13 (all), 63, 66, 72, 86, 99, 101, 105; John Clegg for pages 35, 48, 112, 124, 155, 158, 161, 166, 174, 177, 179, 182; Harry Lacey for pages 136, 148; David Guiterman for pages 65 (both), 71; The Natural History Photographic Agency for page 168 by J. Blossom and page 170 by L. Hugh Newman; and Christine Bousefield for her numerous line drawings, and Reginald Davis and Baz East for additional illustrations.

PREFACE TO
THE SECOND EDITION

This book is a guide to the care of the most commonly kept pet animals. In addition to information on their general management and breeding, as much background information is given about the nature and habits of these very different animals as is relevant to keeping them satisfactorily in captivity.

There has recently been one notable change in the range of animals available to pet owners in Europe. The ever popular Mediterranean tortoises are no longer imported. This is the result of an EEC regulation banning the importation of the three species *T. graeca*, *hermanni* and *marginata* to the member states since 1 January 1984. The ban has stopped the trade in these tortoises for the first time since 1890, and is the result of a long campaign by those who feared the extinction of Mediterranean tortoises.

All who have a genuine interest in such appealing creatures will welcome the restriction, although two unpleasant effects of the ban are already discernible: a 'black market' in the species is reported, together with increased importation of species from other parts of the world, such as the American Carolina Box, and Horsfield's tortoise which is native to the Asian steppes. Further legislation may yet be needed to extend protection to all species.

I have, nevertheless, retained an amended chapter on Mediterranean tortoises for reference purposes, since many remain in the country and, in favourable conditions, should survive for many years to come.

In other ways the range of animals available is largely unchanged and there was no need to alter the text of this edition for that reason. Veterinary details have been updated where necessary, but matters of general information, natural history and care remain much as before. Canaries still feature in mine rescue teams; the dog licence remains at 37½p, unchanged since 1876, despite a government White Paper exploring the possibility of abandoning the national dog licence scheme, and instead allowing local authorities to fix their own, using the proceeds to fund a dog warden scheme.

The need to teach responsible pet ownership also remains quite unchanged and as urgent as before.

Tina Hearne

GLOSSARY

canines	eye-teeth between incisors and molars
carnivore	animal which feeds on flesh
castration	removal of the testicles by surgery
commensal	living close to man to benefit from his food supply
crepuscular	active at dawn and dusk
cull	to reduce a population by selective killing
diurnal	active during the day; daily
exotic	foreign
fancier	hobbyist with a particular interest in improving breeds
feral	reversion to the wild state
gestation	the period of pregnancy
herbivore	animal which feeds on plants
hybrid	animal produced by crossing 2 different species
incisors	front teeth
indigenous	native
in kindle	pregnant
in season	period during which female is receptive to male
lactation	period during which milk is secreted
molars	cheek or back teeth
mutation	biological variation that can be passed on genetically
nocturnal	active at night
oestrous cycle	period of 'season' or 'heat' preceding or coinciding with ovulation
omnivore	animal which feeds on both plant and animal foods
on heat	synonymous with 'in season'
ovulation	release of ova (egg cells) from ovaries
ruminant	animal which chews the cud
tractable	easy to handle

CARNIVORES

The Dog	*Canis familiaris*

The dog is the undisputed leader of the pet world. The first animal domesticated, it has been man's constant companion for 10,000 years. It may even have been a kind of pet—tolerated, or possibly welcomed as a useful scavenger in early camp settlements—before its long history as a working animal.

Varieties There are more than 200 breeds of dog in the world, classified as follows:

SPORTING DOGS

Hounds The hounds are the breeds trained to hunt other animals. They are divided into scent and sight hounds, according to the way they work. Most scent hounds work in packs, and are short-legged, some slow enough to be followed on foot. The scent hounds are Foxhound, Staghound, Otterhound, and the three bred for hare coursing: Basset, Beagle, and Harrier. The Bloodhound, used to track man, works like a sight hound.

The sight—or gaze—hounds are long-legged hounds relying on speed and sight. They hunt independently, or in pairs, and include Afghan, Wolfhound, Deerhound, Elkhound, Borzoi (wolf), Saluki (gazelle and hare), Finnish Spitz (bear), Basenji, Rhodesian Ridgeback (lion), Greyhound (now used more for racing, but originally for hare and deer coursing), Whippet (bred mainly from small Greyhounds for racing), and Dachshund (classed as a hound in this country, although it can do the work of a terrier).

Terriers The terriers are small but courageous hunting dogs which will themselves go to earth to bolt their quarry (rabbit, fox, otter) from its hole. They have enduring value as rat-catchers. Well-known terrier breeds include

Beagle
(hound)

Airedale
(terrier)

English
Springer
Spaniel
(gundog)

Dalmatian
(utility
dog)

Collie and
Shetland
sheepdog (front)
(working dogs)

Pekingese
with puppy
(toydogs)

Airedale, Border, Cairn, Dandie Dinmont, Fox, Irish, Kerry Blue, Lakeland, Scottish, Sealyham, Skye, Welsh, and several others. The Bull Terrier is different. It is a Bulldog × English White Terrier cross, originally bred when dog-fighting was a sport.

Gundogs These are the breeds trained to assist man in hunting game birds and water fowl. Traditionally Setters, Pointers, and Weimaraners are the tracking dogs; Spaniels and Retrievers pick up the shot birds.

Non-sporting Dogs

Utility Dogs These are the breeds not included in the sporting or working categories, yet mostly bred for special tasks. They include Poodles (performing dogs, bred from a water dog, and sometimes trained to retrieve from water), Chow Chow (bred for fur and meat in the East), Schipperke (Belgian barge dog), Keeshond (Dutch barge dog), Dalmatian (carriage escort dog), Bulldog (bull-baiting), and its two descendants the French Bulldog and Boston Terrier, Schnauzer (originally sheep and cattle dogs), Tibetan Terrier (not a terrier at all, but a sheep and watchdog), Tibetan Spaniel (bred in the monasteries, and considered talisman dogs), Tibetan Apsos (sentinel dogs of the Dalai Lama), and the Shih Tzu (palace dogs of Imperial China).

Working Dogs Many of the working dogs are bred and trained to work with sheep and cattle: Corgi, Pyrenean Mountain dog, Shetland Sheepdog, Old English Sheepdog, Collie, Bearded Collie, and Border Collie. Others are the guard dogs of man, his property, and animals: German Shepherd Dog (G.S.D.), Bullmastiff, Rottweiler, Boxer, Great Dane, and Dobermann (all three escort dogs like the Dalmatian, as well as guards). The polar draught breeds fall into this category: Samoyed, Alaskan Malamute, Eskimo, Siberian Husky, and the two famous rescue breeds: St Bernard and Newfoundland (trained to rescue from snow and water respectively).

Toydogs These are the lapdogs, some miniature versions of larger breeds, bred as small house or show dogs. They include Chihuahua, Griffon Bruxellois, Italian

Greyhound, King Charles Spaniel, Cavalier King Charles Spaniel, Maltese, Papillon, Pekingese, Pomeranian, Pug, and Yorkshire Terrier.

Colour The range of colour from black to white includes blue, chestnut, chocolate, cream, fawn, grey, liver, red, sable, sandy, tan, wheaten, and yellow. Dogs may also be bicoloured, e.g. black and white Border Collie; tricoloured, e.g. black, white and tan King Charles Spaniel; brindled (streaked and flecked), e.g. brindle Boston Terrier with white markings; or roan (the presence of white hairs softening the basic coat colour), e.g. roan Cocker Spaniel.

Description The dog, *Canis familiaris*, Linné, is a carnivore (i.e. flesh-eater) of the same family as the wolf and jackal from which it may derive. It is a social, pack animal showing allegiance to the pack leader—an allegiance the domesticated dog transfers to its master.

The dog's teeth are particularly well adapted for its natural life as a hunter. The prominent canines enable it to kill and to hold prey; the presence of flesh teeth (carnassials), which close like scissors, provide a way to slice flesh; and the molars can be used for crushing bones. The dog also has a large stomach that will allow a whole day's food to be eaten at once. If necessary, a dog will bury surplus food and return to it later.

In nature dogs have long legs, tails and faces, and erect, pointed ears. They do not bark. Barking is an acquired habit of the domesticated dog, unknown among wild canines except the coyote. It is now hardly possible to describe a typical dog, since the breeds are more disparate than those of any other domesticated animal. The heavily built Mastiff contrasts with the slim Greyhound and Saluki, and the high-set Airedale with the short-legged Dachshund and Corgi. The weight range is from 1 to 70 kg (Chihuahua; St Bernard); the height at the shoulder from 20 cm to 81 cm (Yorkshire Terrier; Irish Wolfhound).

In some breeds (Greyhound, Borzoi) the head-shape is more elongated than the normal middle-length (G.S.D.); in others (Bulldog, Pekingese) it is much shortened. Coat-type also varies: long-haired (Afghan); medium

(Labrador Retriever); short (Smooth Dachshund); wiry (Scottish Terrier) and several other types (Poodle, Chow Chow).

Evolution   In the absence of proof, authorities fail to agree on the ancestry of the domestic dog.   It may have evolved from the wolf, *Canis lupus*, from the jackal, *Canis aureus*, or from a Siberian wild dog, *Canis poutiatini*, known from fossil remains.   It may even have evolved from hybrids of two or more of these.   The great diversity of size and shape seen in modern breeds supports the theory of more than one single source of origin.

What is much more certain, from archaeological evidence, is that recognizable ancestral types, from which man later created the modern breeds, existed as early as Mesolithic times, when their divergence almost certainly owed nothing to man's interference.

The earliest European dogs were:

Canis inostranzewi   A northern wolf-like dog which gave rise to the large spitz breeds (polar dogs, Elkhound), Mastiff, and St Bernard.
Canis matris optimae   A primitive sheepdog which gave rise to the large herding dogs including the German Shepherd Dog.
Canis intermedius   A medium-sized dog which is believed to be the ancestor of the scent hounds.
Canis palustris   Considered the ancestor of the terriers, Schnauzer, Pinscher, and some of the smaller spitz breeds including the Pomeranian.
Canis leinieri   A large dog with narrow head.   This is the ancestor of the northern sight hounds: Irish Wolfhound, Deerhound, and Borzoi.

The southern sight hounds—Saluki, Greyhound, and Afghan—are not of European descent, and are more likely to derive from the Indian wolf, *Canis pallipes*.   There is good evidence that these breeds were recognizable at the time of the great cultures of the Near East.   The Saluki was a hunting dog in the Sumerian period 5000 BC; the Greyhound appears on an Egyptian tomb carving dating back to 4000 BC; and the Afghan was also known in Egypt in the third millennium BC.

Domestication The dog was first domesticated when man was a hunter and food gatherer. At that time the dog's own hunting skills, particularly its superior turn of speed, more acute hearing, and better scenting powers would have been of the greatest service to man.

It is usually thought, however, that the dog was a companion before a working partner. None of the canines except the largest wolves is a threat to man, nor frightened of him, and all the canines, being scavengers and carrion-eaters, would have been attracted to the earliest camp sites by the presence of food. One can imagine that in times of plenty they might have been allowed to clean up the remains of carcasses, and even have been welcomed for the purpose. Soon puppies would ingratiate themselves into the settlement, and be easily tamed. The realization that a man accompanied by one of these tamed dogs was a far better hunter—and safer from attack by beasts of prey—was the beginning of the remarkable relationship between man and dog that dates back at least 10,000 years.

Suitability as Pets Mongrels and crossbreds usually make delightful pets, and can be more robust and less temperamental than some purebreds. They cost less to buy—but not to keep. When buying a mongrel puppy it should be remembered that the size of its feet is as reliable a guide as any to its future size.

Pedigree dogs were all originally purpose-bred, often for a particular form of work that will render them unsuitable pets for most families. A scent hound such as a Beagle can be difficult to contain once it picks up a trail; a gundog such as an Irish Setter or Weimaraner, bored by having nothing to do all day, will invent activities to satisfy its boisterous nature.

Another problem is that certain physical characteristics, exaggerated in breeding, may predispose a dog to particular disorders. Long-backed, short-legged dogs (Dachshund) are subject to spinal troubles. The breeds with protruding eyes (Boston Terrier, Pekingese) are prone to eye ailments and injuries. The short-headed dogs (Boxer, Pekingese) are noted for respiratory disorders that make them slobber, snore, wheeze, and vomit;

for eczema in the folds of skin on the face; and for overcrowded teeth which collect tartar and tend to fall out prematurely. These breeds are also high-risk patients under anaesthetic, and some (Boston Terrier) high-risk when whelping because the puppy has a broad head, and the dam a small pelvis.

Two of the most important matters to be considered when choosing a pet dog are size and coat-type. Beginners are always advised against the extremes of size. A dog in the range 6 kg–18 kg (Cairn Terrier—Standard Poodle) usually has a stronger constitution and is better suited to the house. Larger breeds up to about 30 kg (Golden Retriever) are generally more tolerant of children, but a dog of 40 kg (G.S.D.) is too large for most houses. The most satisfactory coat-type for a pet is the middle length (Corgi, Labrador Retriever). Short-haired breeds (Smooth Dachshund) have little protection in bad weather; curly-coated dogs (Poodle) need expert clipping; long-haired dogs (Rough-coated Collie) need daily grooming, and are liable to flea infestation and to eczema.

The final consideration is whether to buy a dog or bitch. Most people seem to agree that, provided she can be properly contained while on heat, the bitch makes the better pet, being affectionate, and more amenable to family life.

Life Cycle in Captivity   The young are born after a gestation period of about 63 days (60–65). Litter size varies with the breed. Litters of 1–20 puppies are recorded, but 8–10 is considered large. Normally small breeds may be expected to give birth to 2 or 3 puppies; larger breeds to 5 or 6.

The puppies are born with eyes and ears closed, but with hair. Birth weights are in the range 100–500 g. During the first week of life the puppies can only mew, crawl around blindly, and suckle. In the 2nd week they learn to stand and become much more adventurous. The eyes begin to open about the 10th day, although sight and hearing are not complete before they are a month old. Towards the end of the 3rd week they will begin to take some solid food, including any disgorged for them by the dam. During the 4th week teething begins, and weaning is completed between the 5th and 7th week.

After weaning, the puppies need to stay with the dam until they are 8–12 weeks old. The milk teeth are shed between the 4th and 6th month. Maturity is reached between the 10th and 12th month. Bitches are capable of breeding from the time of their first heat, but should not be mated until the 2nd or 3rd heat. Dogs have a reproductive span from maturity throughout their adult life. Breeders usually retire both sexes from breeding at 8 years. The life span in captivity is generally in the range of 10–16 years, but ages in excess of 16 are common.

Male and Female   The male is known as the dog; the female as the bitch. The bitch has 6–10 teats, arranged in two rows. Since there is an obvious correlation between their number and the usual size of a litter, larger breeds tend to have 8 or more teats. Females are normally smaller than the dogs of their own breed. This example refers to the adult English Setter:

Height at shoulder	Bitch 60–64 cm	Dog 64–68 cm
Weight	Bitch 25–28 kg	Dog 27–30 kg

The difference between bitch and dog puppies is so obvious that sexing them is not a problem, as it can be with kittens or the small rodents.

Breeding   All parental responsibility falls on the dam. On no account should a dog be introduced to a new-born litter he has sired. He will not recognize the puppies as his own offspring, nor remember the bitch, who is liable to attack him as an intruder.

During pregnancy the dam's normal diet should first be improved, and later increased, by the addition of extra milk, meat, fish, and a vitamin-mineral supplement. In the last 3 weeks, when the development of the puppies is greatest, feed the dam 2 meals a day, and allow her to eat as much as she wants, which may be three times her usual amount. The bitch will normally go off food at least for the day her puppies are born.

Prepare a whelping-box two weeks before the litter is due, and try to accustom the bitch to it by encouraging her to sleep there. Three sides should be high enough to exclude draughts, and one low enough for her to step over comfortably. The box has to be large enough for the bitch to be able to turn round easily, and long enough for her to lie stretched on her side when suckling the young. For the whelping it will need a disposable proprietary litter, or torn strips of paper. Afterwards, newspapers may be used, but always covered by a layer of towelling or a blanket. Puppies will lick newspaper, and may be poisoned by the printer's ink.

Whelping   Two veterinary check-ups are needed. One should be carried out about a fortnight before whelping. That is often the convenient time to have the hair of a long-haired breed clipped on the underside of the body. The other should be soon after the birth. At this visit the veterinary surgeon may remove the dewclaws that can cause distress later in life, and should be asked to destroy any unwanted puppies. A bitch is unlikely to rear more than 5 or 6 puppies on her own milk, and unless they are to be hand-reared, surplus puppies must be humanely destroyed.

The bitch becomes increasingly nervous as whelping time approaches, and needs her owner's reassurance, but it is seldom necessary to have a veterinary surgeon present at the birth except for the large-headed breeds (Boxer, Bulldog, Pekingese, Chihuahua). These may have to be delivered by Caesarean section.

The certain sign that whelping is imminent is a drop in the bitch's temperature to below 37·8°C. The first puppy should be born within an hour of the bitch beginning really strong contractions. Longer straining than this is a danger signal, and the veterinary surgeon should be called. It may be some hours before the entire litter is born, for labour is not necessarily continuous, and there may be time for the dam to rest between the birth of the puppies.

Each puppy is born in a separate membraneous sac, trailing an umbilical cord, and followed by the placenta (the afterbirth). It sometimes happens that the sac ruptures during the birth, releasing the puppy, but it is incapable of breaking out unaided. The dam will normally release it, and bite off the umbilical cord. If she does not, the owner must. Break the cord between the thumbnails (a clean cut can cause bleeding), and do not tie it unless there is bleeding. The puppy will probably start to breathe immediately its head is free. If not, place it by the dam who will clean away the mucous and induce breathing by vigorous licking. If the dam continues to ignore the puppy, rub it vigorously with a towel, but first clean away the mucous because the puppy's failure to breathe is almost certainly due to blocked air passages.

There is no need to assist a dam who is dealing with her whelping competently, but she should be under observation. Count the number of placentas, since any retained would set up a uterine infection. It is quite normal for the bitch to eat the placentas.

Pairing Owners wishing to breed from a pedigree bitch are advised to contact the owner of a suitable stud dog. At stud, the kennel staff will supervise the mating. Information about stud kennels can be found in the press, or obtained from the Kennel Club, 1–4 Clarges Street, London, W.1. If possible select a nearby stud to avoid tiring your bitch with a long journey, but do not inbreed her to a closely related dog. Inbreeding is used by professional breeders to stabilize characteristics, but there is always a risk of congenital deformities and undersized pups. It is not a method recommended to novices.

Owners of crossbred or mongrel bitches should make a private arrangement with the owner of a suitable local dog. Before the mating both the dog and bitch should have a veterinary check-up, and the bitch must be wormed if necessary, or she will pass on the infestation to her puppies.

On the day of the mating, take the bitch to the dog's home, never the dog to hers. Introduce them to each other in an enclosed space such as a room or yard, where they will not be distracted. If both are of similar size, and if the time is right, they will mate readily. The bitch is receptive for only a few days during the 2nd week of her season, immediately after her bleeding stops. It is customary to introduce the bitch to the dog on two consecutive days, in order to increase the chance of her becoming pregnant.

Should a misalliance (an accidental mating) occur, it is possible for a veterinary surgeon to give the bitch an injection that will prevent her having pups. The injection prevents the fertilized eggs becoming implanted in the uterus, and must therefore be given as soon as possible after the mating, and preferably within 36 hours.

Never attempt to separate mating dogs. The penis becomes further enlarged after penetrating the bitch, and once the two are locked together in this way it is physically impossible to separate them before the male has ejaculated—perhaps 30 minutes after mounting.

A pedigree bitch is not spoilt for future breeding by giving birth to a litter of mongrel or crossbred puppies. She may even bear a mixed litter of purebreds, mongrels, and crossbreds. This is possible because more than one dog can mate her during her period of heat, siring different puppies.

Oestrous Cycle Normally the bitch comes into season twice a year for about 3 weeks at a time. For these 6 weeks she is attractive to dogs and during the middle week of each season will mate with any, producing 2 litters a year. The dogs locate her by scent, and try as hard to reach her as she will try to escape to them. She must, therefore, be under constant supervision, and very safely contained while on heat. This is such a difficult task that some owners board their bitches for these 6 weeks of the year, or have them injected every 6 months to prevent their coming into season. Others tether the bitch indoors, or keep her in a dog-proof enclosure. A preparation such as veterinary Amplex can be used, with variable results, to mask the scent. Clearly a bitch in season can have only restricted exercise on a lead.

Sometimes a bitch will develop a false or pseudopregnancy about 9 weeks after her season, although she has not been mated. It lasts about 2 months, during which she will show signs of being pregnant. The symptoms—including the production of milk, increased weight, anxiety, and selecting a whelping place—are so convincing that it is very difficult to believe there are no developing pups in the uterus. The condition is thought to be due to a hormonal imbalance, and veterinary advice should be sought. The pseudopregnancies tend to recur, and to lead to uterine infection (pyometra). Spaying is the only certain cure.

Spaying and Neutering The more usual reason for spaying is that of controlling the dog population. A spayed bitch undergoes major surgery, after which she can neither come into season nor become pregnant.

If given the opportunity, a dog will pursue every bitch on heat in the neighbourhood, and will leave home for days on end to seek out and mate stray bitches, adding relentlessly to the stray dog problem in this country. A neutered dog is castrated to render him incapable of breeding, and less liable to roam. It also has the effect of making dogs less aggressive, and is sometimes recommended for uncontrollable, problem dogs. Spaying and neutering render bitches and dogs ineligible for showing.

Rearing Puppies Warmth is of the greatest importance in rearing puppies. It is advisable to allow all bitches to whelp and nurse their puppies indoors. If it should be necessary for the puppies to be outside for the first few weeks of life, great care must be taken to keep them in a high enough temperature, and away from draughts. An outhouse is likely to need heating—at least at night—with an infra-red lamp or a greenhouse heater.

At 3 weeks the puppies can be given cows' milk, and during the 4th week need to be weaned further. Feed minced or scraped raw meat and gravy, milk puddings, soaked puppy meal, and dietary supplements sold under a reliable brand name such as Vetzyme. By the 7th week weaning is complete, and the dam's own milk supply will cease. By the 8th week, when the puppies are mature enough to leave the dam, they are fully independent.

Never allow a puppy to be taken away from the dam, nor accept a new puppy into the household before it is 8 weeks old. As the puppies are weaned, and need to suckle less, the bitch can be given some respite from them. Keep the litter in a low-sided enclosure that allows her to enter and leave at will, while keeping the puppies safe.

Between the ages of 8 weeks and 4 months, the young grow so rapidly that they need 4 meals a day: 2 meals of warm milk and cereal (porridge, Farex, brown bread, puppy meal); 2 meals of meat with puppy meal soaked in gravy. It is very important to include fat in the diet, together with cod liver oil, vitamin B tablets, and a mineral supplement. Avoid rabbit, chicken, and rib bones, but a marrow bone is good. It is not only another source of calcium, but exercises the jaws, helps teething, and keeps the puppy happily occupied for hours.

One milk feed can be omitted at about 4 months, and by 8 months it is sufficient to feed just the 2 meat meals a day. Continue to offer a bowl of milk, but not at the same time as a meat meal. As the puppies mature at 10–12 months they are ready to start the adult feeding programme. Remember that as the number of meals given decreases, the quantity of food given increases.

Freshly-drawn drinking water must be available to the puppies all the time, but not food. Meals need only be left down for 15 minutes or so, and then cleared away. Avoid feeding tit-bits between meals.

A puppy will need veterinary attention several times during the first year of life.: a health check at purchase; probably early treatment for worms; and a programme of vaccination beginning at 8 to 12 weeks against the infectious canine diseases. These are canine distemper (including hardpad), viral hepatitis, parvovirus, adeno-virus, two forms of leptospirosis, and the kennel cough syndrome. Some puppies will also need to have retained

Pair of kennels with runs attached

Traditional dog kennel

milk teeth removed by a veterinary surgeon at about 6 months.

Never allow a puppy to mix with other dogs, or to walk in the street, before it has been inoculated against the infectious diseases. In fact a young puppy needs a lot of rest, and gets all the exercise it needs in the house and garden. Do not walk a puppy before the age of 6 months.

Housing Most pet dogs are expected to be house dogs, sleeping indoors at night. The only special accommodation they need is a good-sized bed of wood, canvas or basket-work, raised off the ground on short legs to avoid draughts, and made comfortable with a cushion and a washable rug or blanket. A dog is particular about the siting of its bed, and will not use it if he dislikes its position, or finds it too small for comfort.

A pet dog expected to sleep outside at night needs a big, draught-proof, weather-proof kennel, but owners find a kennel useful for any dog. Two designs can be recommended. One is the traditional kennel (above) which is usually used for the smaller breeds, and is good if enclosed in a run. The other (p. 24) is suitable for larger breeds, and incorporates a small run.

Kennels should be erected on a concrete apron, raised from the ground to avoid rising damp, and sited to face away from the prevailing wind. The best are constructed of tongued and grooved boarding on a timber frame, with a separate lining. The roof is covered with

Dog walking on lead

roofing felt, and overhangs the walls. The whole front is hinged to open up for easy cleaning.

Plenty of clean, dry straw, wood-wool or torn paper is recommended for the bedding material, covered with a washable rug or blanket. Avoid positioning the bed opposite the opening, where there will be no protection from the weather.

Exercise The need for exercise is absolutely basic to the dog, and is usually the biggest demand he makes on his owner. Most small breeds, and the rather ponderous Bulldog and its descendants, need only about 1·5 km a day, but big mongrels, most hounds, the gundogs, the working dogs, some utility dogs (Dalmatian), and some terriers (Airedale, Border, Fox, Lakeland) will appreciate as much as 16 km a day. Since it is both dangerous and antisocial to allow them to roam at will, they must be exercised on a lead until they reach open ground where they will be safe to run free.

Apart from his daily walks, a dog must be let out first thing every morning to urinate, and needs several hours of complete freedom every day. It is worth improving the fences and gate catches to make the garden—or part of it— dog proof. If this is impossible, then the running chain (p. 27) will keep the dog safe, while allowing a certain amount of exercise. This device is better than tethering a dog on a short chain, which not only denies him exercise,

26

but tends to make him ill-tempered, but by its design allows only repetitive pacing or running to and fro, and so cannot, unfortunately, be recommended unreservedly.

Gauge 12 wire is usually strong enough for the line (but use gauge 8 for big dogs) to which the dog's chain is attached by means of a ring. The line is stretched taut from the dog's kennel to a wall, post, or tree, and should be a little higher than the dog in order to take the weight of the chain off him. There must be a stop on the line (see above) when it is attached to a post or tree to prevent the dog from accidentally entangling himself around it.

Dogs in Cars Most dogs enjoy a car ride, and even those which suffer car sickness can have tranquillizer tablets prescribed. On long journeys carry drinking water, and allow frequent stops for the dog to urinate and take a few minutes' exercise on a lead.

In hot weather it is not safe to leave a dog in a parked car, even one parked in the shade with the windows partly open. Dogs are highly susceptible to heat exhaustion, and as the interior of a stationary car heats up, are liable to collapse.

Dog Licences From the age of 6 months pet dogs must be licensed, and wear an identity disc on the collar in public (max. penalty: £400). The collar will need to be adjusted or replaced occasionally in order to remain comfortable throughout life. The licence is available from the Post Office for a fee of 37½p, and is renewable annually. Note: new legislation is expected soon.

Cleaning A dog's bed blanket and cushion needs shaking or vacuum-cleaning every day, and should be washed frequently. The bed itself will need a periodic clean, particularly basket-work beds which are the most difficult to keep hygienic.

Each day droppings should be picked up from runs, and from the garden, and the bedding in the kennel adjusted for comfort. Straw is best changed once a week when the kennel is swept out. It is unlikely the kennel will need to be scrubbed and disinfected each week, but do so as often as is necessary.

Grooming Grooming a dog adequately is likely to take 10 minutes a day, and the puppy should gradually be accustomed to this as a normal part of daily routine. It is carried out most efficiently, with the least strain on dog and owner, when the puppy has been expected from an early age to stand on a bench or table.

Bath a dog when need be, but not unnecessarily. Use warm water and a dog shampoo, rinse thoroughly, and blow dry with a hair-dryer, or rub vigorously with a towel. Most dogs can properly be given just two or three baths a year, although white dogs at least will need more. Bathing is not advised for puppies under 6 months old, as they are liable to catch cold, nor for wire-haired breeds, since this tends to soften the hair. Dry shampoos are available for use when bathing is impracticable.

Grooming a dog is much more a matter of brushing. Daily brushing improves the circulation and muscle tone, removes tangles, burrs, loose hairs, and dead skin. Long-haired breeds are brushed and then combed, and those with coats expected to stand up (Keeshond, Pomeranian, Chow Chow) are finally brushed backwards, against the way the fur grows. Short-haired breeds are groomed with a soft brush or a hound glove, and wire-haired breeds with a stiff brush and a metal comb.

Owners of breeds that need clipping (Poodles), or trimming (Miniature Schnauzer, rough-haired terriers) are advised to have this expert grooming done professionally.

In the wild state the dog moults twice a year, shedding

the light summer coat in autumn, and the thick winter coat in spring. In the artifical conditions of captivity the moulting is much more prolonged, and may appear continuous. It helps to use a stripping comb, which effectively removes dead hair from the coat, and this is particularly recommended in summer, since it has the effect of thinning the coat.

A dog's claws grow continuously, and will need clipping or not according to how much they are worn down naturally. Have them clipped professionally by a veterinary surgeon or at a dog beauty parlour until you feel confident enough to tackle the task personally. Use animal nail clippers, sold at pet shops, or a small pair of electrician's wire cutters, and as with budgerigars, rabbits, and tortoises, take care to avoid cutting into the blood and nerve supply (see picture, p. 49).

Food and Water   Adult dogs of the larger breeds have big enough stomachs to take the whole day's food at one meal. Smaller breeds more often have their day's food divided into two meals: a meat meal in the evening, and a small breakfast of milk and cereal. In the wild it is safest to bolt the food quickly, taking time only to tear it into pieces small enough to swallow, and then to lie up quietly to digest it. This natural habit is retained in captivity, and is the reason why dogs bolt their food, and why the main meal is fed in the evening, when the dog can rest undisturbed afterwards.

As hunters, wild dogs are primarily carnivorous, but not entirely so. They also take vegetable matter directly, and indirectly as a result of eating the stomach contents of their prey. In captivity the standard diet consists of protein, fat, and carbohydrate, with vitamins and minerals. Safe bones can be given to adult dogs, as to puppies, and a clean dish of fresh drinking water must always be available.

Fresh meat is fed raw, but liver and other offal should be cooked. There is no need to cut the meat except for young puppies, or dogs which have lost their teeth. The meat will contain enough fat for most pet dogs, but not for those expending a lot of energy each day, growing puppies, and pregnant or nursing bitches. When extra

29

fat is required, spoon a little melted fat from the frying pan or grill pan onto the meat. Cereal foods such as dog biscuits or meal, which are fortified with extra vitamins and minerals, and wholemeal bread is fed with the meat in the proportion $\frac{2}{3}$ meat to $\frac{1}{3}$ cereal. A few suitable table scraps—meat, vegetables, gravy, fat, boned fish, egg, cheese, brown or wholemeal bread, breakfast cereal—may also be fed, and at some other time during the day offer a dish of milk.

Alternatively, a dog may be fed a proprietary brand of dog food, following the manufacturer's recommendations. Dehydrated foods must be fed with plenty of water; tins of meat need cereal added; and whole diets, whether tinned, moist, or dried, are fed alone. The careful balance of ingredients is lost if these complete diets are fed with extra carbohydrates. These proprietary foods are considered good, but check that the dog has a bone or large dog biscuit to exercise his teeth and jaws, to counteract the effect of eating only soft food.

There is much variation in the amount eaten by individual dogs and by different breeds, but in general small dogs need more food for their weight than larger ones. A Maltese of 4 kg needs perhaps $\frac{1}{2}$ can a day; a Mastiff of 64 kg probably 5 cans (but not 8 as is suggested by its being 16 times heavier).

Handling and Restraint To lift a small puppy, scoop it up with one hand under the chest; to lift a dog, stand sideways to it, and take the weight of the chest and hindquarters in both arms.

If it is ever necessary to restrain a dog from biting, for instance when clipping its claws, use the tape muzzle illustrated here, which is a 1-m length of bandage, crossed beneath the chin and tied behind the ears. Remove the tape immediately the treatment is finished.

Training A puppy is old enough to be house-trained from the age of 8 weeks. It is simply a matter of putting him outside to urinate first thing in the morning, and every other time he wakes from sleep, and putting him out after every meal and drink. He quickly understands what is expected of him if he is praised when he urinates outside, but is scolded when inside. A watchful owner,

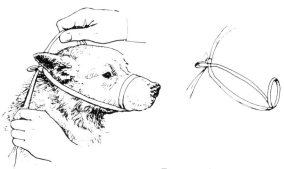

Tape muzzle

who anticipates well for the puppy, will accomplish house-training within a month.

Advanced obedience training is inappropriate for pet dogs, but they should learn to walk to heel on a lead, to come when called, and to understand the commands 'sit' and 'over'. They should also be taught not to jump up, not to bark without good reason, not to chase people, vehicles, or farm animals, and not to worry at table.

Informal training towards good manners begins from the day the puppy joins the household, because he understands his owner's attitude and tone of voice, and is eager to please, but formal training of commands should not begin before he is 6 months old. 'Sit' is the easiest to teach. As you say 'sit', hold the collar with one hand, and push down the puppy's hindquarters with the other. The puppy will learn the meaning of 'over' just as effortlessly if you always give the 'sit' command at the kerb, and then say 'over' as you begin to cross the road.

'Heel' is taught while walking a puppy on his lead. When he pulls forward to overtake you, or drags behind, gently jerk him into position with the lead. Limit walking on a lead to no more than 5 minutes at a time until the puppy has learned to walk to heel at your speed. Sometimes a check chain (choke chain) is used to teach this command to big or boisterous dogs, but it can be both cruel and dangerous if used roughly or fitted incorrectly.

Check chain correctly fastened

It should be fastened so that the collar loosens immediately the tension on the lead slackens.

'Come' is taught by holding the puppy on a 6-m cord, attached to its collar, while it plays. Call the puppy by name, adding the command 'come', and at the same time gathering in the cord. Praise success, and repeat the lesson about six times a day, gradually lengthening the cord. When the time is right, test your teaching by allowing the puppy to play free in an enclosed space. He should now come immediately he is called, but if he just runs off continue to call the same command, and praise him if he does eventually surrender to it. Even belated obedience must be praised during training.

The best owners and trainers always handle dogs quietly and unobtrusively. To this end it is advisable to use hand signals as well as words during training so that the dog can eventually be controlled silently. This is an invaluable means of communication with a dog which goes deaf in old age, and can be learned, with plenty more besides, at a local dog-training class which novice owners are strongly advised to join.

Ailments RABIES is the most feared dog disease, and transmissible to man. It does not yet occur in Britain, and in an effort to keep it out all dogs entering the country must spend 6 months at a quarantine kennel.

The serious INFECTIOUS CANINE DISEASES (page 24) should pose no threat, since dogs can be protected by inoculation, although boosters are needed. Combination vaccines are available which give protection against several of the diseases simultaneously.

Dampness, draughts, and cold may lead to RESPIRATORY TRACT INFECTIONS, so dogs should be rubbed down when they come in wet, and kept in dry accommodation, free from extremes of temperature. PANTING does not indicate troubled breathing, but heat. The dog sweats only through the pads of his feet and uses panting as a means of cooling down in hot weather.

To correct a minor digestive upset a dog will eat grass as an emetic, but PERSISTENT VOMITING should receive veterinary attention. A dog which drags itself along the ground is probably trying to relieve the irritation of BLOCKED ANAL GLANDS. The condition causes intense discomfort and needs veterinary treatment.

Dogs—or more often puppies—may contract the fungus infection RINGWORM, which is transmissible to man. Take veterinary advice if round, dry, scaly, bald patches appear on the skin. Veterinary advice must also be sought if there is any sign of the EXTERNAL PARASITES that can torment dogs: fleas, lice, ticks, and mites. When a dog shakes its head and scratches its ears it is probably because of an infestation of mites, or the presence of wax or grass seeds. Prompt veterinary attention is required. Fleas and lice can be treated at home with a proprietary powder, and the bedding washed or renewed to destroy flea eggs and larvae.

INTERNAL PARASITES in the form of several kinds of worm can live in the dog's alimentary canal. Some, notably *Toxocara canis*, are transmissible to man. Veterinary, and not home, treatment is necessary for worms, very often once a year.

The Cat   *Felis catus*

The cat is kept by some for its sheer grace and beauty; by others as a mouser; perhaps by most for its companionable yet undemanding presence about the house. It is always bracketed with the dog as pre-eminent among household pets, yet in its history its popularity has fluctuated from one extreme to the other.

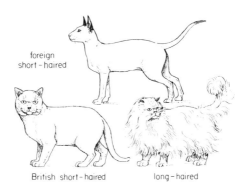

foreign
short-haired

British short-haired   long-haired

Varieties Although most cats are mongrel, there are
more than 100 breeds, classified into three main
types. (Not all the breeds mentioned here are recognized
by the Governing Council of the Cat Fancy of Great
Britain.)

Foreign Short-haired Cats This group of about 35
breeds includes some of the most instantly recognizable:
the Siamese, Burmese, Abyssinian, Russian Blue, and the
curly-coated Rex cats. All have a characteristic lean,
long-limbed look, with a wedge-shaped face, and pointed
ears and tails.

British Short-haired Cats This group of about 20
breeds is characterized by a more rounded outline, with a
compact body, shorter legs, and broader head. It includes the tailless Manx, and the Scottish Fold with
drooping ears, but most of the breeds are known simply by
their colour. Colours include brown, red, and silver
tabby; tortoiseshell, and tortoiseshell and white; blue-
eyed white, orange-eyed white, and odd-eyed white;
black, blue, and cream; blue-cream; and bicoloured.

Long-haired Cats These are the cats whose coats are
their glory. Long-haired cats were originally called
Angoras, and later Persians. Over 50 breeds have been
developed, including long-haired varieties of the colours
listed above for British short-haired cats. All of these
long-hairs have the characteristic thickset, short-legged
outline, with a broad, snub-nosed head, and a short but

34

very full tail. This cobby appearance is emphasized by the length of the coat, and by the ruff which, for show purposes, is brushed up to frame the face. Colourpoints (known as Himalayans in America) combine the outline of the typical long-haired cat with the markings of the Siamese.

However, not all long-haired breeds display the cobby outline. The Balinese is a long-haired Siamese, retaining the form of the short-haired Siamese; the Somali is a long-haired Abyssinian, retaining the form of the short-haired Abyssinian; and the Cymric, as yet an unrecognized breed, is a long-haired Manx.

Colour The range of self (i.e. single) colours is black, brown (several shades including havana and chocolate), ruddy, red, blue, lilac, cream, and white.

Many cats are two-coloured. Those with big patches of a self colour on a white ground—most often black on white—are known in Britain as bicoloured; in the USA as parti-coloured. Blue and cream are intermingled to produce the blue-cream coat. Abyssinians, Chinchillas,

Kitten practising hunting skills

Siamese

bicoloured
black and white

tortoiseshell

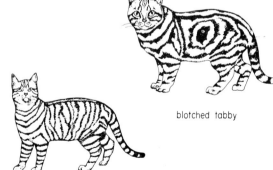

blotched tabby

striped tabby

and smoke varieties have light coats ticked with a darker colour. Spotted cats have dark spots on a paler ground. Tabbies have darker markings on a brown, red, or silver ground. All but new-born Siamese, which show no markings, have darker coloured 'points' (ears, feet, legs, tail) and face mask on a paler ground.

Tortoiseshell cats are three-coloured; tortoiseshell and white cats are four-coloured.

Kittens are born blue-eyed, but adult blue-eyed whites are nearly always deaf. Other eye colours are green, hazel, yellow, orange, and copper. Odd-eyed whites have one blue, and one orange eye.

Description　The cat, *Felis catus*, Linné, is of the same family as the tiger, lion, puma, leopard, ocelot, and lynx. Like the dog, the domesticated cat is a carnivore, yet the differences between them may seem more obvious than the similarities. The dog is a social, pack animal; the cat is solitary and detached, with an independence that makes it the most self-sufficient of pet mammals.

The cat family tends to have coats marked with darker stripes or spots on a paler ground, and the tabbies, which are nearest in form to their wild ancestors, also have such a coat. The original tabby coat is the striped (mackerel) pattern; the blotched tabby is thought to be a later mutation. The patterns always remain separate, but are persistent. Most modern cats, even if self-coloured, show traces of one or other of the patterns, at least when they are kittens.

Another characteristic of this family is the short muzzle. The domesticated cat has short jaws with only 30 teeth, whereas the dog has 42. The cat's teeth are highly specialized, adapted only for tearing and slicing flesh. The powerful canines will kill and hold prey, and the blade-like molars of the lower jaw close scissor-fashion within those of the upper jaw to slice flesh. The incisors are inconspicuous, there is no lateral movement of the jaws to allow for grinding, and the molars have no grinding surface. The molars cannot be used effectively for crushing bones, but the cat has a rasp-like tongue covered with papillae that allow it to scrape the flesh from bones and aid grooming.

Cats are by nature nocturnal. During the day much of the time is spent sleeping and resting, dawn and dusk being the times of major activity. It is often said that cats can see in the dark, which is not true, but they are able to see in poor light. The pupils of the eyes dilate as the light decreases in order to take advantage of the slightest glimmer which is amplified by a mirror-like membrane, the *tapetum*, common to nocturnal mammals. Conversely, in bright light the pupils narrow to a vertical slit.

When hunting, the cat first stalks or lies in wait for its prey, and then closes in at high speed. The canines and the claws can be used to kill prey once caught. The retractable claws are normally sheathed, so the cat is able to keep them very sharp for use as a weapon.

Although it can reach very high speeds over short distances, the cat has small lungs, and so lacks the stamina to maintain speed like the dog. It is compensated by having remarkable agility. Its ability to climb and to leap is not only an aid to hunting, but also the means by which it very often escapes predators such as the fox.

Domestication Although wild cats are widely distributed throughout the world, man first used weasels, polecats, and ferrets as mousers. There is evidence that the cat was not domesticated until the 2nd millennium BC in Egypt, and was virtually unknown as a domestic animal in Europe before Roman times.

In Egypt the cat achieved the status of a sacred animal, and enjoyed great protection under the law. The penalty for deliberately killing a cat was death, and there were fines for accidental killings. When a cat died a natural death the body was embalmed, and the whole family shaved off their eyebrows as a sign of mourning.

Although the Egyptians are known to have guarded their cats jealously, inevitably some were smuggled out of Egypt, and others went feral and spread outside North Africa to interbreed with other wild cats. It is thought that the modern European cats are derived from three wild species, all capable of interbreeding: *Felis lybica*, the North African cat; *Felis manul*, the wild cat of the Asian steppe; and *Felis silvestris*, the European wild cat.

As a domesticated animal in Europe the cat's fortunes declined. It became strangely identified with the devil and witchcraft, and black cats in particular were persecuted. This sorry state of affairs persisted as long as witch-hunting, but gradually during the 18th, and certainly the 19th century, cats began to regain their popularity. The first cat show, held in 1871, symbolized their new status.

Life Cycle in Captivity The young are born after a gestation period of 63 days (61–69). Litters of 1–9 kittens are recorded, but 3–5 is more usual.

The kittens are born with ears and eyes closed, but with hair. Birth weights are in the range 90–140 g, increasing by 80–100 g per week. During the 1st week of life the young are able to suckle, to crawl, and to mew, but warm, well-fed kittens are quiet and sleep much of the time. In the 2nd week the eyes open (by the 10th day) and the incisors begin to show. Towards the end of the 3rd week the kittens should be taught to lap milk, skimmed or powdered, from the fingers, and from the 4th week the mother's milk will need to be supplemented if the weight gain is to be maintained. All the milk teeth are cut by the 5th week, and weaning is complete in the 7th or 8th week.

After weaning the young are best left with the mother until they are at least 8 weeks old. The permanent teeth are cut between the 4th and 7th months. Males reach maturity at about 6 months; females usually earlier—sometimes as early as 5 months—but they are not full grown, and are too young to be subjected to the strain of breeding before 12 months. Most breeders retire females by the age of 8 years, and males by the age of 10 years. Life span in captivity is normally in the range 12–16 years, although ages in excess of 25 years have been recorded.

Male and Female The entire male is known as the tom; the unspayed female as the queen. The males may be expected to be larger than the females, average adult weights being in the range 2·25–3 kg for females; 3·5–5·9 kg for males. Neutered cats tend to put on extra weight. The females have 8 teats, and toms may develop thick necks and big cheek muscles, giving a heavy-jowled

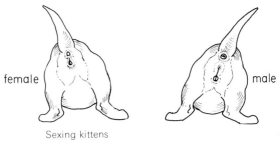

female   male

Sexing kittens

look. Male kittens have a round genital opening; female kittens a genital slit. The distance between the anus and genital opening is greater in the male, to allow for the descent of the testicles. True tortoiseshells are invariably female; orange cats are more likely to be male.

Breeding   The queen rears her litter alone, with no assistance from the tom cat. The kittens are normally safe with the tom only when they reach 4 or 5 months. Before that he is liable to attack and kill them.

The pregnant queen needs no special attention during the first 5 weeks of pregnancy, but she should not be handled needlessly. During the last 4 weeks, when her weight gain shows, her appetite will increase, and she will need extra food such as more meat and fish. During the final 2 weeks of pregnancy the development of the kittens is very rapid, and the queen should be allowed as much milk, water, meat, and fish as she will take. A daily vitamin tablet, and a dietary supplement are beneficial.

Although queens are notorious for shunning the box provided by their owner, a box should be prepared for the birth as for the bitch (p. 19), although positioned in a darker place. Kittens are susceptible to bright light even before their eyelids open.

Kittening   Kittening is similar to whelping, the kittens also being born in separate membraneous sacs, and the information given in the dog section (pp. 20-1) is applicable to cats.

Pairing   Pairing takes place most often at night, and is usually successful. The female is stimulated to 'call' by male presence. She ovulates only in response to copulation with the result that there are always sperms available to fertilize the egg-cells. The pairing of pedigree cats takes place at breeding catteries. Suitable addresses can be found in the press. The pairing of mongrel cats is unfortunately almost always accidental.

Oestrous Cycle There is much variation between breeds and individual cats, but all queens come into season repeatedly twice a year until mated. Each season may last from 3 to 14 days. During this time the queens are restlessly searching for a mate, and may cry continually. Indeed, when a cat is in season, she is said to be 'calling'.

Spaying and Neutering There is already a stray cat population in Britain of 1,000,000 or more. Since it is all but impossible to control the breeding of pet cats, there is a responsibility on owners to have them spayed or neutered between the ages of 5 and 6 months (see dogs p. 23). Unspayed queens are likely to give birth to 2 litters every year, and entire toms are generally considered to be unsuitable household pets. They have an especially disturbing howl, known as a 'caterwaul' and a habit of marking out their territory by spraying urine in the house.

Rearing Kittens This is similar to rearing puppies and the information on pp. 23–5 applies here.

The queen will herself house-train her kittens if she is supplied with a good dirt-tray. It must be kept very

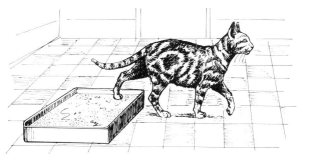

Dirt-tray for indoor use

Cat-door

clean or she will refuse to use it, and the layer of litter—proprietary cat litter, sand, or dry earth—must be deep enough for her to teach the kittens to bury their droppings.

Like puppies, kittens will need veterinary attention during their first year, particularly vaccination against feline infectious enteritis and feline influenza.

Kittens can safely be taken from the dam at 8–12 weeks. At this age they are fully weaned, and well able to fend for themselves.

Housing Cats need little more accommodation than a warm, dry bed similar to that provided for the dog (p. 25). They are particularly susceptible to draughts, dampness, and cold, so the bed should be raised about 10 cm off the floor for comfort.

Although nocturnal by nature, it is advisable to keep pet cats in at night. Outside they run the risks of bad weather, fighting, road accidents, indiscriminate breeding, cat-thieves, poisoning, teasing, and air-gun pellets. Cats soon adapt to a more diurnal existence if accustomed to it from an early age, and if given a dirt-tray at night. Older cats cannot be expected to change the habits of a lifetime, but should have access to a good bed in a shed, or be able to enter the house by way of a cat-door.

Exercise Cats do not need a great deal of exercise, but they do need freedom. Once they have been spayed or

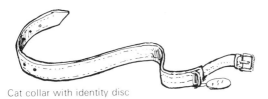

Cat collar with identity disc

neutered, most pet cats are allowed their freedom. When this is undesirable, certain breeds such as the Siamese will walk on a lead.

Cat collars—if used—need to be made of elastic, about 1 cm wide, or to have an elastic inset. There is a danger that any collar may act as a noose if the cat becomes caught up in a tree or on railings.

All cats must have access to a scratching post. They are fastidious creatures, and their routine grooming includes keeping the claws clean and sharpened. Cats confined indoors much of the time will need to be given a scratching post, or a proprietary scratching block.

In addition to taking more obvious exercise, cats—and particularly kittens—need to play. Suitable toys are ping-pong balls, balls of paper, wooden cotton reels, and paper cartons. Avoid toys that may be swallowed or, if broken, will leave sharp edges.

Grooming Ideally, short-haired cats should be groomed daily, but if they resent that much attention—and some do—they will almost certainly be able to keep themselves in good condition except when moulting. Cats normally moult twice a year, and unless groomed by the owner will swallow loose hair. This hair will mat together with food to form a hairball in the throat or intestine, causing a very serious blockage.

Long-haired breeds must be groomed daily, or even twice daily. Unless a long-haired coat has such attention, its hair will tangle so badly that the knots will have to be cut out. It is essential to accustom a long-haired cat to grooming when it is very young, being careful not to overtax its patience in the beginning. First remove burrs, fragments of dead leaves, small twigs, and so on, and then use a coarse comb, a finer comb, and a brush—in that order.

43

It is unlikely that a cat will need washing, although some owners shampoo white ones. In some circumstances a damp sponge may be used, or there are dry shampoos on the market.

Food and Water The adult cat needs 2 meals a day. Like the dog, the cat in captivity needs protein, fat, carbohydrate, with vitamins and minerals, milk and water, and safe bones. Water is important, and should be put down twice a day, although many cats prefer to drink from a dripping tap or puddle. Unlike the dog, the cat cannot manufacture vitamin C and so benefits from a teaspoonful of chopped greenstuff scattered on each meal.

The information given in the dog section (pp. 29–30) is applicable, except in quantities. A rough guide is that a cat may be expected to eat 1/30th of its weight each day.

Unless accustomed to a varied diet from an early age, cats tend to develop food fads as a result of which they may be eating an unbalanced diet. It is essential to feed a variety of foods from an early age.

Scratching post

Handling and Restraint   To lift a small kitten, scoop it up with one hand under the chest; to lift a cat, place one hand under the chest, and support the weight of the hindquarters with the other hand.

A cat can be effectively restrained from scratching and biting, for instance when injured in an accident, by being wrapped in a towel or blanket.

Ailments   Like dogs, cats entering this country are quarantined by law for 6 months in an attempt to prevent the introduction of RABIES.

The most virulent cat diseases are FELINE INFECTIOUS ENTERITIS and FELINE INFLUENZA.   Both are very serious diseases from which kittens need protection by inoculation at 10 weeks, with a further injection at 12 weeks. Immunity will last about a year, but a veterinary surgeon will advise how often boosters are needed.

In addition there are several infectious RESPIRATORY TRACT diseases, and cats may suffer from BRONCHITIS or PNEUMONIA.   Most make good recoveries if nursed carefully and given prompt veterinary treatment.

HAIR-BALLS and DIGESTIVE complaints are often treated by the cat itself.   All cats use grass as an emetic, and town cats should have some grown in a flower pot if need be.   Persistent digestive ailments need veterinary attention.

SKIN DISEASES of cats are similar to skin diseases of the dog, and veterinary advice should be sought.   Similarly EAR 'CANKER' may be due to one of many causes, including the presence of a mite.   Whatever the cause, cats showing symptoms of discomfort about the ears should be taken to a veterinary surgeon.   FLEAS and LICE may be successfully treated at home, using a proprietary powder.   Clean the bedding to eradicate flea eggs and larvae.   INTERNAL PARASITES need veterinary treatment.

If a cat or kitten has to be destroyed, it should be put down by a veterinary surgeon, or by an animal welfare organization such as the RSPCA.

LAGOMORPHS AND RODENTS

The Rabbit	*Oryctolagus cuniculus*

Rabbits were traditionally bred for their meat and for their fur—known as 'coney'. The practice of keeping them solely as pets is comparatively modern, but the rabbit has an established reputation for being one of the most agreeable of all: attractive, undemanding, and hardy.

Description Rabbits have been domesticated for so long that there is now inevitable variation from their wild ancestors, but all except the lop-eared rabbits retain the three most instantly recognizable features of the wild parent stock: the long, erect ears; the long hind legs; and the short, distinctive scut, or tail. The middle-sized Dutch and English, with their well-defined markings, typify the strains very often chosen as pets.

Varieties There are some 35 breeds, perhaps best summarized under the three groupings used for show purposes:

Normal Breeds The normal breeds have a coat of short, dense fur (underhair) interspersed with longer, stiffer, guard hairs. Examples include the Argentés, the Beverens, the Californian, the Chinchillas, and the New Zealands.

Rex and Satin Breeds These breeds are distinguished by their magnificent velvet or satin-like coats in which the guard hairs are absent, or do not stand up above the level of the underhair. The Rexes appeared first, and are derived from the Castorrex, a French mutation that was first shown in Britain in 1927. The Satin, with its roll-back fur, was imported from the USA in 1947.

Fancy Breeds The fancy breeds are the pre-eminent show rabbits, and were not necessarily bred for meat or fur. All are remarkable in some way: the Angoras for their 'wool'; the Lop for the great length of the ears; the

Himalayan

Dutch

Polish and Netherland Dwarfs for their small size; the Flemish Giant for its great size; the Belgian Hare for its similarity to that animal (although pure rabbit); the English, Himalayan, and Dutch for their body markings.

The colours of domesticated rabbits include black, white, blue, lilac, fawn, chocolate, tan, havana, and several shades of grey. Some are self-coloured; some have 'broken' colours, e.g. blue and tan, or tortoiseshell; others have 'ticking' produced by guard hairs, e.g. the Argenté Champagne, which has so much white ticking among the blue-black underhair that the overall effect is silver.

Origins Although introduced to most of their Empire by the Romans, there is no written record of the rabbit in this country before the 12th century. There is, for instance, no mention in the Domesday Book of 1086, although rabbits were probably brought here soon after that by the Normans. Their value as a fur animal—which became apparent with selective breeding—was not recognized until they were domesticated by monks in the 16th century. Their value as a meat animal, however, was recognized immediately, and for centuries they were 'farmed' in enclosed warrens, where they were protected by strict game laws. As late as 1816 a new Act was passed with a maximum penalty of 7 years' transportation. However, the rabbit's popularity declined towards the end of the 19th century as it became established in the wild in numbers that posed a threat to an increasingly arable

Small or middle-sized rabbits make the best pets for children

form of agriculture. It was at this time, as the wild rabbit gained its reputation as a pest, that the serious selective breeding of our modern rabbit varieties began.

Lagomorphs and Rodents Rabbits (together with the hares) are lagomorphs. The rabbit, *Oryctolagus cuniculus* (Linné), was for many years classified as a rodent, and although the two orders are superficially similar, it is now known from fossil evidence that as far back as the Eocene period (50–70 million years ago) the two orders were distinct, and probably evolved from different ancestors. Lagomorphs differ from the rodents in several minor respects. Characteristically they have long, erect ears; hind legs that are longer than the fore; and an inability to hold food in the forelimbs when feeding. Two of the major differences relate to their teeth, and to their habit of refection. While rodents have 4 incisor teeth, lagomorphs have 6. In the upper jaw they are paired—smaller, subsidiary ones behind larger front ones. This arrangement gave rise to the old name of Duplicidentata i.e. double-toothed rodents. The lagomorphs also move the mouth from side to side when chewing, unlike rodents, since the lower jaw is hinged differently.

The habit of refection—also known as coprophagy and pseudorumination—enables the lagomorphs to pass food twice through the stomach for better assimilation. The ruminant mammals, who live on a similarly indigestible diet, achieve this re-ingestion by chewing the cud. There is a biblical reference to the coney and the hare 'that cheweth the cud' (Leviticus XI 5– 6), but the lagomorphs do not regurgitate food from the stomach into the mouth. Instead they produce two kinds of faeces: the normal dry faecal pellet, and a softer pellet that is not evacuated from the body. The soft pellets are taken from the anus by the lagomorph when it is at rest, and swallowed whole without being chewed. As domesticated rabbits have become diurnal, they re-ingest food this way at night; with wild rabbits, refection takes place during the day. This habit is in no way a behavioural abnormality, but an essential part of the digestive process of lagomorphs. Those deprived of practising it under laboratory conditions have suffered a serious vitamin B deficiency, and have died within a few weeks.

Natural Habitat   The rabbit is indigenous only to the Iberian Peninsula, and it may have spread naturally across the Pyrenees into France. Its present world-wide distribution—and consequent threat to agriculture—is due entirely to man. Its spread was particularly marked in Roman times, and again in the last century when it was unwisely introduced to the Australasian and South American continents.

blood supply

cut here

Rabbit skull and overgrown claw

49

Adaptations to Habitats   The wild rabbit has proved itself one of the most adaptable mammals on earth.   Its successes following introduction to Australia, New Zealand, Argentina, and Tierra del Fuego—where it was subject to little or no predation—have passed into history. It thrives best on temperate grasslands, but can live wherever grass can grow, in areas of high rainfall or low. In the wild it normally follows a crepuscular or nocturnal habit, grazing pasture, arable land, or market-garden crops when safe, and lying up during the day—in a burrow if the sub-soil is suitable for excavation.

The burrowing habit is normally of great importance to the wild rabbit.   It encourages big populations to live close together so that danger signals—for instance the well-known thumping of the hind feet—are quickly transmitted, providing an efficient early warning system against attack.   It also provides protection from predators, from dehydration, and from the elements, and allows the young to be reared in the relative safety of a nursery burrow.

Although burrowing is usually one of the key factors of its success, the rabbit has recently shown that it is not essential.   Some rabbits survived myxomatosis in Britain by living above ground, where they were relatively safe from the carrier of the disease—the rabbit flea—which cannot breed so freely when exposed to surface temperatures.   In abandoning the normally safe burrow, those rabbits again demonstrated the tremendous adaptability of the species.

Life Cycle in Captivity   The gestation period varies within the limits of 28–34 days, but is normally reckoned to be 31 days plus 12 hours for a larger breed.   The young are known as kittens, and are usually born at night. Litters tend to be large, averaging about 5 young, but with 10 not uncommon.   Birth weight varies very much according to the size of the breed, and to the number of young in the litter.   It may be 30 g or less for one of the small breeds, or for a kitten from a large litter; 70 g or more for one of the largest breeds, or for a kitten from a litter of only 2 young.

The kittens are born helpless with no fur except a light

down, and with eyes and ears closed. By the end of the first week fur is beginning to grow; by the end of the second week the eyes open (approx. 10th day) and also the ears (approx. 12th day). During the third week the young begin to leave the nest (16–18 days) and to take some solid food. The mother will suckle her young for 6 weeks, and possibly as long as 8 weeks. Small breeds may reach sexual maturity by 4 months; larger breeds may take 6 months or longer. It is advisable to delay breeding until 6 or 10 months respectively. The male's active reproductive life extends from 3 years for small breeds to 4 years for larger breeds, although he may remain fertile beyond this. It is usually recommended that females be retired from breeding at 2 or 3 years, similarly depending on breed-size. Life span in captivity averages 6–7 years, but ages in excess of 12 years have been recorded.

Male and Female   The male is known as the buck; the female as the doe. Adult weights vary with breed from 0·9 kg to 5·5 kg or more. When sexing young rabbits, it should be noted that the distance between the anus and the genital opening is greater in the male. The male has a round genital opening; the female a slit that may be V-shaped. The doe rabbit has 6 teats. With maturity, the scrotal sac of the male becomes obvious, as do the 2 main secondary sexual characteristics, viz. the larger body size of the doe, and the broader head of the buck.

Breeding   Most pet-keepers choose to keep either a single buck, or a pair of does. However, if homes can be found for the plentiful offspring of their rabbits, owners are likely to be successful in breeding them.

All parental responsibility falls on the doe, who should be segregated from the buck immediately after mating. She needs tranquillity, a hutch of her own, and an improved diet. Gradually increase the amount of milk, mash, and cereal given so that, by the time the young are born, she is receiving twice her normal amount of feed. By the time the young are 3 weeks old, and taking a little solid food themselves, she may well need 3 times her normal amount of feed. Both pregnancy and lactation are very demanding, and the doe cannot rear her litter without this increase in her own diet.

One feature of a professional breeding hutch that can be incorporated into a pet doe's hutch at breeding time is a shelf. When the litter is born this simple fixture will allow the doe to get away from the kittens from time to time, and will give her some much needed respite from the demands of her young family. Towards the end of her pregnancy put plenty of chopped hay in the hutch. A few days before giving birth to her young, the doe constructs a nest, lined with her own fur, in the sleeping compartment.

It sometimes happens that a doe who cannot produce enough milk to suckle her young will cull their numbers or even kill the entire litter. This also happens when the kittens are disturbed by over-anxious owners. Experienced breeders will tempt the doe from the nest with a titbit, in order to gain time to have a quick look at the litter on the 2nd day, and to remove any still-born kittens. The inexperienced are warned that it is very chancy to have any contact with the kittens before they emerge from the nest voluntarily.

At weaning the doe needs to be separated from the young to recuperate. It is unfair to expect a doe to raise more than 3 litters a year. When serious breeders need more kittens than this from a particular doe, they spare her the strain of lactation by fostering the kittens to other does who have litters of similar age. Although fostering would not normally be considered by the pet-keeper, it is a possible way to rear young orphan rabbits. It is usually successful if begun before the young are 3 weeks old, and if the foster doe's kittens are of similar age.

Pairing To effect a mating the doe should be introduced to the buck's hutch, never he to the doe's, for she is liable to attack him on her own territory. On neutral ground the buck will take some time to settle, possibly showing more interest in his new surroundings than in the doe. Mating should take place within a few minutes if she is receptive and exhibits lordosis, the characteristic rigid stance of a receptive female with curved spine and raised tail. If not receptive, she may be mated to another buck, or re-introduced on another occasion.

Oestrous Cycle The remarkable fecundity of rabbits is partly due to the fact that the doe does not have a true oestrous cycle, but is able to conceive at almost any time, even immediately after giving birth. If conditions are right she will remain on heat for very long periods. Certain factors such as lactation, underfeeding, moulting, and the seasons will limit this potential, but rabbits can, in theory, be mated all through the year. In practice, however, it is found that fertile matings are most often effected between January and June, and that October, November, and December are the worst months for breeding.

Because there is no oestrous cycle the doe does not ovulate regularly, but in response to the stimulus of the buck. This characteristic can lead to the phenomenon of 'pseudopregnancy' when a sexually excited doe—who may not have been in actual contact with a male—nevertheless will develop the symptoms of pregnancy, including the production of milk. Pseudopregnancy usually lasts about 17 days until it is terminated by nest-building. It may be suspected if nest-building coincides with the 17th day after an incident when the doe could have been aroused. Nest-building in a true pregnancy would not begin so early.

Rearing Juveniles The young are usually segregated at weaning to prevent the mating of very young animals. Young males must be housed singly from the age of 3 months, when they would begin to fight seriously.

Housing An adequate hutch is of great importance to a rabbit's well-being. The responsible owner will spare no effort to provide a hutch of good size, design, construction, and finish. Minimum size for a pair of medium-sized does is 150 cm × 60 cm × 60 cm high. As bought hutches are very often too small and too insubstantial, a home-made one may well be superior.

The traditional design shown on p. 54, if well built, is as good as any. Important features are the separate day and night compartments; an overhanging, pitched roof to direct rain and snow away from the interior; a sturdy tongued and grooved construction on a wooden frame; a smooth interior finish to resist gnawing; a weldmesh or wiremesh front to the day compartment for light and ventilation; safe catches; and legs that lift the hutch well

Traditional rabbit hutch with additional shutter for use at night

clear of the ground, away from rising damp and from vermin.

The finish is similarly important. A tarred felt roof is recommended for good weatherproofing, and the exterior walls can be covered with the same material. Alternative finishes for the exterior walls are gloss paint, creosote, or varnish. Formica would be a suitable material for the hutch lining. Wood, if used for the interior, should not be painted, but given a coat of polyurethane varnish. The interior walls of the hutch should cover the framework and present a completely smooth surface to the rabbit. Any projection would allow the rabbit to gnaw, and the fabric of the hutch would soon be destroyed.

The rabbit must, however, be allowed to gnaw on something, to reduce the length of the incisor teeth. A large, bark-covered log in the day compartment will best serve this purpose. There is no danger of rabbits or rodents swallowing splinters of wood. The illustration on p. 49 shows the wide gap, the diastema, that separates the incisors from the molar teeth. Because of this gap, and its hare lip, the rabbit is able to close the mouth behind the incisors when not eating, isolating anything being gnawed from the mouth.

The flooring of the hutch should be strong and level,

with no front ledge to hinder cleaning.   As it will have to withstand urine, a wooden floor should be given a coat of polyurethane varnish or, better, be protected by a galvanized or aluminium tray.   A plastic tray may be used if it is incorporated into the hutch in such a way that the rabbit cannot get to its edges to begin gnawing. Throughout the hutch there should always be a layer of wood chippings, cat litter, or peatmoss litter, 5 cm deep, to absorb urine.   In the sleeping compartment this is covered with plenty of hay or straw, the quantity of bedding being increased as the temperature falls.

Temperature   The preferred temperature range is from 10°C to 18°C, but rabbits are tolerant of a much wider range of temperature.   In this country they can safely be housed in outside hutches throughout the year, providing the hutches are so built and positioned that the rabbits are protected from the extremes of temperature. In winter added protection can be given by using a shutter or throw-over cover on the hutch at night, and by moving the hutch into the shelter of a lean-to or outhouse.   In winter the hutch must be sited in a sheltered position, facing away from the prevailing westerly winds and from cold northerly airstreams.   Facing S or SE would usually be right.   In summer a S-facing hutch would overheat, but a SE-facing position is again recommended.   The critical upper temperature for rabbits is 28°C, at which they are liable to suffer heat exhaustion.   If this temperature occurs, rabbits are best released into the garden.   If they must still be caged, the hutch needs to be moved into the shade and its roof hosed down with cold water, and then propped partly open to increase the ventilation.   In high temperatures survival can be dependent on improved ventilation.

Exercise   Rabbits are by nature active animals which jump, climb, and burrow.   One of the owner's first tasks is to make provision for their exercise.   For preference rabbits should have the freedom of a safely bounded garden.   This freedom not only gives ample opportunity for exercise, but stimulates the development of personality.   The connection between animal personality and environment is noticeable in all pets, but in none more

Rabbits in fenced enclosure

than the rabbit. A close-caged rabbit, deprived of a stimulating environment, is all too often recognized by its marked lethargy.

Owners may find it convenient to fence off an area around the hutch in such a way that the rabbit can, by using a ramp or flight of steps, use it freely during the day. A fence 1 m high will safely contain the rabbit if it is also sunk into the ground to prevent escape by burrowing. In addition, most owners find the portable outdoor run a satisfactory way of giving limited exercise and fresh grazing every day. Some rabbits will share a run peaceably with other animals such as cavies and tortoises.

The run is constructed of wiremesh on a triangular wooden frame, and is an adaptation of a very old grazing hutch. About $\frac{1}{4}$ of the length is roofed and may be enclosed as a shelter. The triangular construction gives size and stability without excessive weight, and this run is difficult to overturn. It is probably easier to move every day than the more clumsy rectangular design sometimes seen. Normally these runs need no mesh on the bottom. As a rule, tame rabbits are ineffective burrowers, and would not have time to tunnel their way out of a run that is constantly being moved.

Portable outdoor run

Rectangular outdoor run

57

(*Note* These runs are unsuitable for use as a rabbit's permanent living quarters, although they are sometimes used as such in summer. If so, the enclosed section must be floored 5 cm above the ground, and fitted with a shelf for sleeping. They are, however, vulnerable to attack by vermin, and liable to become damp.)

For a rabbit, toys are no substitute for contact with the natural world. Two playthings, nevertheless, are usually appreciated: a fallen tree trunk for climbing, jumping, and gnawing; and a drainpipe of large diameter for use as a substitute burrow. In wet weather, it helps if rabbits can be exercised indoors, or in a shed or outhouse. Close supervision is necessary because of their tendency to gnaw, but an effort should be made to give them a change of scene even in bad weather.

Cleaning Twice a week—more if necessary—the rabbits have to be transferred to temporary accommodation while their own is cleaned. This is a simple chore if the hutch is at table height, and designed to open up for easy access. The floor is scraped clean, using a three-sided scraper in the corners, and all the soiled floor-litter and bedding collected in a bucket or wheelbarrow ready for burning. If necessary, the hutch is scrubbed with detergent water, rinsed thoroughly, and allowed to dry before replenishing the litter and bedding, and returning the rabbits. Depending on how many rabbits use the hutch, and on how much time they spend in it, it may be necessary to change the floor-litter more frequently than this—perhaps every other day. It is not usually necessary to scrub more than once a week to keep the hutch hygienic, and bedding should be disturbed as little as possible.

Grooming A clean hutch and good food are the prerequisites for a rabbit's lustrous coat. Even show rabbits may have no more grooming than a rub with a silk handkerchief. Rabbits are fastidious about their own grooming, and providing they are housed in clean accommodation, most breeds are able to keep themselves in good condition. Even so, many owners like to brush once a week with a small hairbrush. The brushing itself may not be really necessary, but it is important to remove burrs or small pieces of twig that may be adhering to the coat, and to observe the rabbit's condition. Pay particular attention to eyes, ears, nose, and anus, and to the claws.

Feeding Rabbits are entirely herbivorous, living only on plant food. They are, therefore, particularly at risk if fed just scraps and greens. Their diet should be a careful balance of cereals, green food, roots, and hay. A good daily maintenance diet for a non-breeding rabbit of medium size would be:

    50 g oats or wheat
    50 g cereal-based mash
    170 g fresh vegetable matter
    85 g hay

(These amounts are increased considerably for 'working' rabbits, i.e. breeding or show rabbits.)

The cereals most often given are oats and wheat, oats being considered better for rabbits. Wheat can be given in the form of wholemeal bread, dampened with milk into a mash, or alternatively baked hard in the oven.

One nutritious mash, much favoured by breeders, is mixed dry in the proportions:

    40% maize meal
    30% crushed oats
    10% bran
    10% fish meal
    10% linseed cake meal

This is made up with hot water, or hot milk, to a crumbly consistency, and served in an earthenware pot that can be scalded each time it is used. Cooked potato, or cooked potato peelings can occasionally be added to the mash. As it soon turns sour, the mash should not be left in the hutch longer than an hour or two.

Most schools and laboratories, and some breeders, find it convenient to feed pellet food, but it is not recommended for pet rabbits. Pellets may meet the nutritional needs of the rabbit, but they give no variety, and no occupation. Because herbivores, by nature, spend much of the time feeding, whole food is preferable.

59

Hay is particularly valuable for keeping rabbits occupied, and so nutritious that it should always be available. Breeders maintain that clover hay is best, but sweet meadow hay is almost as good. In the interests of hygiene, it is best offered in a hay rack, as is green food.

Vegetable food should be washed and dried before being fed to rabbits, and frosted greens discarded. Rabbits with access to good grazing may need very little extra greenstuff; those which have to spend much of the day in a hutch will need about 170 g a day. Feed the natural succession of vegetables and fruits as they become available, including apple, beetroot, Brussels sprouts, cabbage, carrots, cauliflower, celery, chicory, kale, lettuce, parsnip, pears, peas and pea pods, spinach, swede, and turnip. Lettuce contains traces of a poison (lactucarium) and should be fed only in moderation. In addition to garden produce, many wild plants can be fed to rabbits, including agrimony, chickweed, clover, coltsfoot, comfrey, cow parsnip, dandelion, goosegrass, groundsel, hedge parsley, knapweed, plantain, shepherd's purse, and sow thistle. Avoid those plants listed on p. 187.

Rabbits have a particular need for salt in their diet. Without it they rapidly lose condition, both growth-rate and milk-yield being adversely affected. A block of rock salt, and a rabbit mineral lick should be suspended in the hutch permanently.

Being creatures of habit, rabbits need to be fed at regular times. Many owners feed 2 meals a day: a main morning meal of cereal and mash, and an evening meal of vegetable matter. Others prefer to feed 3 meals a day: cereal in the morning; vegetable matter at noon; and an evening mash. Remove uneaten vegetables and greenstuff daily.

Drinking Water Fresh drinking water must always be available. A large-size drip-feed bottle will be needed, since rabbits sometimes drink large amounts. If provided in a heavy pot, water is liable to be spilt or fouled.

Handling Rabbits soon lose their initial nervousness and become tame if handled with confidence from the time of weaning. They are docile by nature, and seldom bite. They should never be picked up by the ears alone, nor by the scruff of the neck. When lifting, always support the whole body weight with one hand beneath the rump. The hand around the base of the ears, or on the scruff of the neck, is there only to control the rabbit's movements. Once in the arms, the rabbit may allow itself to be cradled, or will settle with the head over one shoulder. Although rabbits may kick out and inflict accidental scratches, many of these can be avoided by returning a pet to its hutch backwards.

Ailments   Rabbits have very poor recuperative powers, and failure to seek prompt veterinary advice when they display symptoms of disease, or lose their normal alertness, is often fatal.

The fearful rabbit disease, MYXOMATOSIS, is a severe virus infection spread quickly among wild—and occasionally commercial—rabbit populations by the rabbit flea. Few pet rabbits are likely to be in close proximity to other rabbits to be at risk, but veterinary advice about vaccination should be sought if a pet is considered to be in danger.

EAR MANGE, or ear canker, is an infection caused by a small mite. The early symptom is an irritation that causes the rabbit to scratch the ears, and to shake the head. A powdery brown matter may be visible in the ear. Patent anti-mange preparations are available, and likely to be effective if used at this stage. The rabbit will suffer considerable pain if the condition is allowed to deteriorate. At the more advanced stage there is inflammation and even ulceration, and professional treatment is vital.

An infection equivalent to the common cold in man, with similar symptoms, is known in the rabbit world as SNUFFLES. Apart from being very infectious, there is a real risk of its developing into PNEUMONIA. Isolate the rabbit from others, and consult a veterinary surgeon.

Rabbits react badly to sudden changes in their food, and CONSTIPATION or DIARRHOEA may be due to a simple dietary upset. Immediately increase (for constipation) or decrease (for diarrhoea) the greenstuff given, but seek veterinary advice if diarrhoea persists.

Diarrhoea is a non-specific symptom of more serious

rabbit disease, for instance, COCCIDIOSIS. Coccidiosis is the result of infestation by internal parasites that affect the intestine or liver. Young rabbits are particularly at risk. Apart from persistent diarrhoea, there is a weight loss with the development of a pot-bellied appearance, and exhaustion that leads to death. The rabbit may be saved by prompt veterinary treatment.

The bacilli that cause SCHMORLS DISEASE, a disease of the skin and mucous membranes in rabbits, gain entry by way of injuries to the skin. Bathe wounds with a mild antiseptic, and separate rabbits which are inclined to fight together.

OVERGROWN INCISORS will kill a rabbit by preventing feeding. The length of teeth can be reduced by a veterinary surgeon, but must afterwards be kept in trim by constant gnawing on a log (p. 54). Similarly OVER-GROWN CLAWS, once trimmed, can be kept down if the outdoor run is sometimes put on an abrasive surface, e.g. concrete. Claws can be cut by a veterinary surgeon, or by a careful owner who will take care to avoid cutting into the blood supply. The picture on p. 49 shows where to cut, but in the event of a mishap, bathe with a mild disinfectant.

It is extremely difficult to kill a rabbit cleanly by physical means. If it is necessary to have a rabbit destroyed, it must be taken to a veterinary surgeon who will inject an overdose of anaesthetic.

The Cavy or Guinea-pig   *Cavia porcellus*

Origins The cavy—or guinea-pig—was brought to Europe from the Guiana coast of South America by sailors in the 16th century, and is thought to have been introduced to England by 1750. There are about 20 species, all native to South America, but occurring in areas that are very diverse geographically. *Cavia porcellus* (Pallas 1766) is thought to have derived from the restless cavy of Peru, *Cavia cutleri*, which was domesticated by the Incas as a food animal long before the Spanish conquest. Although some species burrow, this is a grassland animal of the lower slopes of the Andes, and is

Long-haired (Peruvian) cavy

perhaps more likely to seek shelter in rocks, caves, the abandoned burrows of other animals, and in long or tussocky grass. Certainly in captivity its descendant, *Cavia porcellus*, does not seem to know the technique of burrowing, but confidently bulldozes its way through long grass, loose hay, and thick vegetation. It also exhibits another sign of the surface-dweller—that of bearing mature young (cf. hares) able to withstand adverse conditions from birth.

Common Name There is no adequate explanation for the cavy's common name of guinea-pig. 'Guinea' may well have been a corruption of the less well-known 'Guiana', but although explanations for the use of the name 'pig' abound—including allusions to the animal's taste, hair, voice, and so on—none is convincing. Modern usage seems to favour the more scientific 'cavy'.

Varieties Cavies are classified by coat-type into three varieties:

Smooth-haired (English) The most common of the three is the smooth-haired English, which is also known as the American or Bolivian cavy.

Rough-haired (Abyssinian)   The Abyssinian has characteristic rosettes of coarse hair all over the body, and a ridge of hair along the spine.

Long-haired (Peruvian)   The Peruvian has very long, soft hair, and is obviously a variety for the show-bench rather than the pet-keeper.

Since each is *Cavia porcellus*, all three varieties can interbreed, and English/Abyssinian crosses are frequently seen.

Colours   The colours recognized by the cavy fancy at present are white, black, red, cream, beige, golden, chocolate, and lilac.   Single-coloured cavies are known as 'selfs'; bicoloured or tricoloured cavies are said to be 'marked'.   The most common marked (i.e. patterned) cavy is the Dutch, which has a pattern of one or two colours on a white ground, reminiscent of the Dutch rabbit.   The Himalayan cavy is also white with coloured points—nose, ears, and feet—like a Himalayan rabbit or a Siamese cat.   The tortoiseshell is a bicolour, chequered red and black; the tortoiseshell and white is a tricolour chequered red, black and white.   The other marked

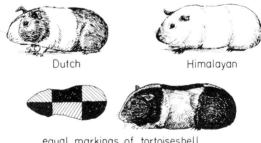

Dutch   Himalayan

equal markings of tortoiseshell

tortoiseshell and white   agouti

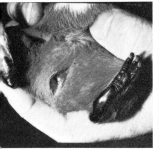
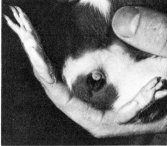

Female cavy   Male cavy

cavies are the agoutis, with golden, silver, cinnamon, or salmon ticking. In these cavies there is no pattern, but instead each hair shades from black to gold, from black to grey, from black to cinnamon, or from black to salmon, giving a distinctive mottled appearance.

Description Cavies are nervous creatures, yet popular both as laboratory animals and as pets because they are particularly easy to tame, to handle, and to control. The heavy, short-legged body inhibits movement and renders them the least agile of all the pet rodents. They run quite fast, but do not climb, and many do not jump, although those given freedom will bound and frisk in long grass with an abandon that surprises those who have seen only the restricted, stereotyped movements of closely confined cavies. The front paws are used for grooming, but it is not characteristic for cavies to use them for feeding, as do most other rodents.

Male and Female The male is known as the boar; the female as the sow. Individuals vary considerably, but the distinctive cobby, tail-less body of an adult male would measure perhaps 25 cm and weigh 1000 g; that of the female would be about 5 cm shorter and weigh possibly 850 g. Both sexes have 2 teats, but can be sexed by the shape of the genital opening: the boar has a round opening, and gentle pressure with the thumb will extrude the penis; the sow has a Y-shaped slit.

Abyssinian sow with litter

Breeding For breeding purposes it is usual to keep one boar with several females. Cavies are capable of reproducing throughout the year, but most often mate in the period from April to September. The females come into season for up to 15 hours at a time, approximately twice a month (every 16–19 days). They also come into season immediately after giving birth, and for this reason the boar should be removed when a sow has her litter to prevent another pregnancy following immediately. The boar is not, however, likely to interfere with the young who are so well-developed at birth, nor are live-born young at risk from their own mother or from other sows in the colony.

The gestation period—averaging 63 days—may seem very long for a rodent, but cavies belong to a sub-order, the Hystricomorpha, that characteristically has exceptionally long pregnancies. The sow does not construct a nest, and the young are born fully haired, with eyes open and milk teeth cut, able to move about. They can even take a little solid food at 2 days. Litters are rather small, but well-managed cavies may be expected to raise litters averaging 3 young. Birth weights vary considerably: single births may produce young weighing 150 g, but weights of 90 g for males, and 85 g for females are satisfactory. Those weighing less than 50 g at birth may fail to survive. The young are suckled by the sow for only 3 weeks, but should not be removed from her until they are 4 or 5 weeks old.

Although sexually mature at 2 months they are much too young for breeding, which should be delayed until they reach at least 6 months. Segregation of the sexes is necessary from the age of 2 months to avoid unwanted pregnancies. Cavies, however, are not prolific breeders: a pet sow should be allowed to raise only 2 litters a year for the 2 years or so she is in breeding condition. Life span in captivity is not likely to exceed 8 years, and is frequently in the range of 4–5 years.

Housing By nature social animals, cavies are best not kept singly. Although in the confines of captivity males would fight, two females will live together peaceably. A wooden rabbit hutch (pp. 53–5) measuring 120 cm × 60 cm × 50 cm high makes a suitable home for two or three cavies. Providing it is properly constructed, well-positioned, and furnished with plenty of dry bedding, healthy cavies are hardy enough to survive most of the winter out of doors in the south of Britain, although they will need the extra protection of an outhouse or lean-to in the north, and in severe winters. Protection from summer heat is just as important as protection from winter cold.

Temperature The preferred temperature range is 16°–20°C. Lower temperatures can be survived under the shelter of abundant bedding although the young do not grow well at temperatures below 13°C. At temperatures higher than 20°C, there is a risk of heat exhaustion, and death may occur at about 32°C—a temperature that could often be reached in a badly-placed hutch on a hot summer day in this country. On such a day it is safest to give the cavies the run of the garden, where they are free to seek a cool refuge under shady vegetation.

Exercise Pet cavies need space and freedom. Every effort should be made to fence the garden—or part of it (pp. 56–7)—or failing that to build an outdoor run

Indoor cavy run

(pp. 56–7) which, if need be, can be shared with friendly rabbits or tortoises.   Obviously a much lower fence than that needed for rabbits will keep cavies safe, and unless there is a local danger of marauding dogs, an outdoor run for cavies needs no mesh on the bottom. They will not attempt to burrow, and mesh may cause soreness about the mouth and hocks.

Indoors, for instance in a classroom, a cavy pen will need a solid base covered with formica or polyurethane varnish, and is best mounted on castors for easy movement.   Unless the cavies are free to enter their own hutch whenever they want, by way of a ramp, one corner of the pen should be covered in as a shelter.

Cleaning   (See also rabbit section p. 58).   The hutches may require daily cleaning if the cavies are confined to them for long periods; twice a week is sufficient if they spend most of their day in the garden, or in an exercise run.

Grooming   All varieties should be groomed with a small brush of suitable size; a toothbrush is big enough for an Abyssinian.   In each case the hair is brushed in the way it grows: from the head downwards on the English variety; upwards and out from the centre of each rosette

68

and from the base of the ridges on an Abyssinian; outwards from a long centre-back parting on a Peruvian so that (for the show-bench) nothing can be seen of the body outline. Unless a Peruvian is to be exhibited, it is kinder—and easier—to trim the hair, particularly the fringe over the eyes.

Food Cavies are entirely herbivorous in captivity, as in the wild, and they have two particular dietary needs: one is for a high daily intake of vitamin C since they, like man, can neither manufacture their own, nor store it in the body; the other is for a daily supply of sweet meadow hay. Hay is so important to them (p. 71) that it is false economy to buy any but the best, and even that has to be picked over to remove harsh plants such as thistle, and the seed-heads of dock.

A satisfactory diet consists of fresh vegetable matter, hay, a cereal-based mash or pellets, and water. The most natural way to feed cavies is to allow them the freedom to graze grass (beware of using herbicides), and to browse on low-growing vegetation. This can be supplemented, as necessary, with all manner of fresh green food including cabbage, kale and lettuce; prunings of ash, elm and fruit trees; roots such as carrot, turnip and swede; fruit as available including apple and pear; and cooked vegetables and vegetable peelings, preferably mixed with bran. Avoid the plants listed on p. 187.

By choice, cavies spend a great part of each day feeding, and this basic diet of fresh vegetable matter, together with a daily bundle of hay, will keep both free and caged cavies well occupied. In addition they need a bowl of cereal-based mash each day: either (a) wholemeal bread soaked in hot or cold milk; or (b) crushed oats and bran soaked in hot water and squeezed to a crumbly consistency.

Useful supplements to add to the mash for pregnant sows, and in winter, include crushed or dissolved vitamin C tablets; cod liver oil for vitamins A and D; and the B group vitamins in the form of Bemax or yeast tablets. The amount of mash eaten by a cavy is an indication of its appetite, and ought to be adjusted accordingly to avoid waste. Any mash remaining after a meal must be removed, as it quickly becomes sour. Dishes must be

69

Drip-feed bottle

washed each time they are used. It is also normal practice, when feeding a caged cavy, to offer green food and hay in a rack, to avoid soiling under foot.

In place of the mash, some breeders and schools, and most laboratories feed pellet food for convenience, but the pet owner is not likely to have recourse to pellet food.

Water Cavies vary considerably in how much they drink, but clean water must always be available, preferably from a drip-feed bottle. Never refuse to give water just because the cavies have access to green food. If pellet food is fed, it is essential to give water.

Handling If gently and regularly handled, cavies become tame and tractable. They very rarely bite. One hand is placed around the back of the neck, and the animal's weight lifted with the free hand. Careless handling of pregnant females in the last weeks of their pregnancy is thought to be one cause of still-births.

Ailments OVERGROWN INCISORS result from the cavy's not being able to gnaw to reduce their length. Have the teeth trimmed professionally, and ensure the animal has access to a bark-covered log. If not worn down naturally on a hard surface, CLAWS too may become overgrown, but may be trimmed like a rabbit's.

With age, some cavies have difficulty in expelling FAECES from the anus. Gently expel them manually, using a tissue. Check that the cavy has exercise, plenty of hay, and other roughage in the diet.

Cavies may fall victim to the deficiency disease SCURVY

70

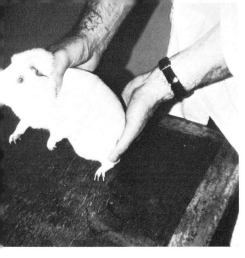

Handling cavy
correctly

if their high vitamin C requirement is not met. Scurvy can also develop when overgrown incisors make feeding so difficult that the vitamin C intake drops, in spite of source foods being available.

Although its importance in the diet is not fully explained, lack of hay is a serious deficiency. Youngsters so deprived quickly develop overgrown incisors, and are at risk from scurvy or general malnutrition; adults so deprived develop the BEHAVIOURAL ABNORMALITIES of stripping themselves, their young, and other cavies bare of hair, and also of eating their bedding.

To treat LICE, bathe the cavy in about 5 cm of tepid water, supporting the body with one hand, and applying a medicated shampoo with the other. Rinse, and towel dry, then spray with a pyrethrum aerosol. INTERNAL PARASITES need veterinary treatment.

Respiratory tract infections such as BRONCHITIS and PNEUMONIA are usually the result of subjecting a cavy to damp or draughts, to which they are particularly susceptible. Prompt veterinary help is essential. In fact, cavies lose condition so rapidly that an owner's failure to act swiftly in response to any ailment may have fatal results.

The Golden Hamster *Mesocricetus auratus*

The first desert rodent to gain popularity as a pet was the golden hamster. It was introduced to Britain in 1931 for laboratory work, and in spite of its nocturnal habit had, by 1945, achieved widespread recognition as a highly successful pet animal.

Description It is appealing, easily tamed, clean, and simple to care for. The body is compact and low to the ground; the tail remarkably short; the ears large, becoming hairless with age. Although the eyes are a particularly fine feature, eyesight is poor, and the golden hamster has real need of its sensitive whiskers.

A good specimen has a short coat of soft, dense fur that appears golden brown above and white on the underside, the juveniles being a little lighter in colour than the adults. A closer look, however, shows the coat to be made up of hair of different lengths: short grey under-hairs, and longer guard hairs that counter shade to grey at their base. This grey hair is noticeable when the fur is

Golden hamster

parted, but does not normally show through. Some of the guard hairs are black or black-tipped, but there should not be enough of these to interfere with the overall golden brown appearance of the hamster. Face markings take the form of white crescents beneath black cheek flashes, and the chest band is an unbroken golden brown.

Varieties   There are over 30 colour varieties of which the National Hamster Council recognizes most for show purposes. Some of the most popular colours are cream, fawn, cinnamon, white, sepia, honey, silverblue, and dark, normal, or light golden. Many can be banded with white to produce, for instance, a 'Cinnamon Band'; some can be bred with black, red, or ruby eyes. There are also Piebalds, mottled with white spots on the ground colour; Mosaics, with brown or cinnamon patches on the ground colour; Satinized hamsters, with fine velvet-like coats; and Tortoiseshells.

Origins   In spite of the problems associated with inbreeding, all the early laboratory animals were reared from just three litter mates, taken from the wild. The story of their capture is well known. In 1930, the Department of Zoology of the Hebrew University at Jerusalem sent an expedition to Syria to collect specimens of an entirely different hamster, *Cricetulus phaeus*, urgently needed for laboratory work. This expedition also found a burrow of golden hamsters: a female and her seven young, which were taken back to the University. There, four escaped and one female was killed by a male, leaving only one male and two females to breed—but this was done so successfully that, it is claimed on good authority, for many years all the laboratory golden hamsters in Europe and America were descendants of the original trio.

Golden or Syrian Hamster   The golden hamster was first recognized as a species in 1839 by Waterhouse, who named it *Mesocricetus auratus*, and brought specimens to London from Aleppo in Syria. Nearly a century later, the 1930 expedition also found its specimens near Aleppo.

This coincidence has gradually given rise to a controversial demand that, now the many colour mutations make a misnomer of the name 'golden', the term 'Syrian' hamster should be used as the common name, following the practice of using geographical names for the two other well-known hamsters, the Chinese hamster, *Cricetulus griseus*, and the European hamster, *Cricetus cricetus*. Geographically, however, the suggestion is not very sound. *Mesocricetus auratus* is neither confined to Syria, nor is it the only hamster found there, and so the ambiguity of referring to a 'Cream Golden hamster' is still preferred by many people.

Natural Habitat The golden hamster is known to occur in Israel as well as Syria, but its probable territory extends farther into Asia Minor. It is a true desert rodent, able to survive the typical hot desert climate: a long, dry summer with a July average temperature as high as 32°C, followed by a short, mild winter with the temperature of no month averaging less than 6°C. Rain falls only in the winter months, and amounts are very small. Damascus has only 218 mm a year. Even so, it is mistaken to think that hot desert rodents have only heat and aridity to contend with. Average temperatures conceal the cold of the desert: in winter snow falls on high ground, and in summer the temperature can fall dramatically at dusk, for with little or no cloud cover to provide insulation, the heat radiates out into space. This is a characteristic of all hot deserts (the greatest diurnal range ever recorded being in Libya: 52°C to −3°C in 24 hours), and results in the formation of heavy dews that provide one of the main sources of water for the animals of the desert.

Adaptations to Natural Habitat As a rule, desert mammals avoid the stresses of their environment more than they adapt to them. They follow a nocturnal, or possibly crepuscular habit, relying on their burrows for protection from heat, cold, dehydration and predators. In this the golden hamster is typical. In summer it uses its burrows as a cool daytime retreat where the relative humidity is perhaps 5 times that of the outside air, and at dusk ventures out, protected by its dense fur coat, to collect food for the storage chamber. The habit of hoarding, which enables the hamster to survive in times of

Gland on
hamster's hip

food scarcity, also gives rise to its common name which is
taken from the German verb meaning to hoard. The
golden hamster has evolved cheek pouches for carrying its
food. These pouches are certainly the most interesting
biological adaptation to a desert habitat, but the hamsters
are not unique in possessing them. Certain of the
American rodents, including pocket mice, *Perognathus*,
and spiny pocket mice, *Liomys*, also have cheek pouches.

Solitary Animals It cannot be recommended that
adults be kept together. Golden hamsters are by
nature solitary animals. In the natural habitat the male
seeks out the female solely for mating. It is thought he
finds her by his sense of smell, and stays with her for only a
short time before returning to his own extensive network
of burrows. These hamsters have a pair of sebaceous
glands that show as a dark spot on each hip. It is not
known how much the secretion of these scent glands is
used for purposes such as the marking of territory by
golden hamsters in the wild, but it must aid a solitary
animal to find a mate. This is almost certainly the main
function of the glands, for they disappear once breeding
age has passed.

Life Cycle in Captivity Young golden hamsters are
born after a gestation period of only 16 days—the shortest
time recorded for any mammal. The average litter has 6
young, although large litters of about 10 are common, and
some with as many as 17 young have been recorded.
Birth weight is ordinarily 2 g, the young being born

75

hairless, with closed eyes and ears, although the incisor teeth show. Rapid development of the pups depends on the mother's ability to feed them adequately. If she is well fed and contented, her young will each weigh 40 g at the age of 4 weeks. By the end of the first week (7th day) fur begins to grow. During the second week (10th day) the ears open, the young leave the nest, begin to take some solid food, and their eyelids open (12th day). The mother will suckle her young for about 3 weeks—at least 20 days, and frequently as long as 25 days. Young females take 7 weeks to reach sexual maturity; the males sometimes take only 4 weeks. The female's reproductive life extends to 12 months, and the male's to 15 months. Life span in captivity averages 2 to 3 years, the males usually outliving the females.

Male and Female Average adult weights are between 100 g and 150 g, the females being slightly larger than the males. The male's body tapers towards the tail; the female's, with the larger pelvis, is more rounded. The sebaceous glands on the hips are much more obvious on the male. Most female golden hamsters bear 14 nipples arranged in 2 rows. Even a young female can be identified by these. They show if the animal is cupped in the hand and her underhair gently blown aside. Additionally, in young females the anus and genital opening are very close; in young males there is a greater distance between.

Breeding As most owners keep only one at a time, they seldom consider breeding from their golden hamster. This is fortunate, for it is not an easy subject for the amateur. Golden hamsters are difficult to pair, and breeding from any but well-bred, healthy animals can result in a rapid deterioration of stock. In addition, the sexes need segregating into two colonies when they are 27 days old, and each hamster will need its own cage soon after that.

Golden hamsters are not good family animals. In captivity, as in the wild, the male needs to return to his own quarters immediately after mating, and he will take no further part in the life of his own family. All parental responsibility falls on the mother. A nursery cage should be prepared for her on the 12th day of her pregnancy. She will resent interference after that.

There should be a generous supply of soft meadow hay from which she can construct a nest to suit the temperature: open-topped in warm surroundings; covered over in colder conditions. After birth the nest should not be disturbed for 14 days, nor the young handled when they explore the cage. An unflurried changing of the food and water, and cleaning of the damp corner following the usual routine will do no harm, and on the 14th day the young family may be lifted into a box while their own cage is thoroughly cleaned. The mother is, however, nervous enough to turn against her young if unduly disturbed before they are 14 days old and will, in any case, cull her litter to remove weaklings and to reduce it to a manageable size.

Only a golden hamster in good condition is capable of rearing a big litter. She quickly becomes exhausted by repeated pregnancies, and should not be expected to bear more than 3 litters in her lifetime. If bred too young, too old, or too often golden hamsters will bear only poor quality litters, or litters they have difficulty in rearing successfully. The females need at least a month's respite between litters, and the males will not sire good quality pups unless they too are adequately rested.

Pairing   Although capable of breeding from 12 weeks to 12 months, the best breeding age has been shown to be 5 to 6 months. Pairing golden hamsters is as difficult as pairing Mongolian gerbils, and the procedure is similar. They are best paired on neutral ground, but if a cage has to be used the female should be introduced to the male's cage, never the male to the female's. 20 minutes is sufficient time for successful mating, but too long for an unreceptive female. She is fiercely territorial and if not in season is likely to chase and fight the intruding male—to the death unless he is rescued by his owner. For this reason the pair should never be left unattended. It is as well to supervise them, wearing stout gloves. If they begin to fight the hamsters must be separated, although it is not easy to part them.

Oestrous Cycle   The females come into season for at

least 4 hours every 4 days throughout the year, and immediately after giving birth. Being nocturnal animals they are likely to be receptive to a male only at night—not much before 10 p.m. Even experienced breeders do not know a sure way of telling when a female is in season. The normal practice is to introduce the pair to each other on as many as 5 consecutive nights, until they are seen to pair. If unsuccessful, it is worth trying another partner. A female who finds one male incompatible may mate readily with another. If, when introduced to the male, the female turns away from him and busies herself with diversions such as pouching food, it is best to part them before she becomes aggressive. A receptive female faces the male and 'freezes' into a characteristic posture known as lordosis, with curved spine, raised tail, and a fixed stare. It is quite normal for much social grooming to take place during the mating session.

Colony Rearing of Juveniles All the pups should be taken from the nursery cage by the 27th day to rest the mother, and to prevent a sexually mature 28-day-old from attempting to mate her. The young should be segregated into two groups according to sex, and may be housed in colonies for a while until fighting begins, but each hamster will need housing singly at about 5 weeks. Golden hamsters may seem tractable enough as pets, but with their own kind they can be most pugnacious. In the home fighting would normally be noticed, but colony-reared juveniles in schools need inspecting every day.

Wooden
box cage

The danger there is that juveniles who sleep peacefully together during the day may be inflicting bites on each other in night-time fighting.

Housing   As so many hamsters would sleep all day out of sight if allowed to burrow, most are housed in cages. Wood is the recommended building material, metal being cold to the touch, and plastic dangerous if chewed. There is a wide variety of commercial cages available, although the expense of production means many are smaller than one would wish.   For this reason a well-designed home-made cage may have an advantage over a bought one.   If hardwood is too expensive to use, softwood lined with Formica offcuts is a possible alternative.   When making drawings for a cage it is important to allow plenty of space, and to design a cage that is well ventilated but not draughty.   A wooden, box-type cage fitted with a removable glass front and a separate galleried sleeping compartment, and ventilated by a wire-mesh lid is a simple, but effective design.   It allows the hamster full freedom of movement within the space available, and privacy and warmth in the sleeping compartment, which needs plenty of bedding.

  Since a pet hamster is likely to spend its entire life alone, it is particularly important to provide it with as large and interesting a home environment as possible.   In the cage, ramps, ladders, galleries, logs, wooden cotton reels and nuts all give scope for play, and opportunity to wear down the growing incisor teeth.   To protect the fabric of the cage from this gnawing, it should be as smooth as possible inside, with a lining to protect the framework.   Exposed ledges are irresistible to a rodent.   It is, nevertheless, essential to allow the golden hamster to gnaw.   In metal cages, with nothing else available, hamsters deprived of wood will spend hours each evening trying to trim their incisors on the roof bars.

  The daytime sleepiness of golden hamsters belies their true nature.   At night they are highly active, and stories of hamster escapes are legion.   It is wise to fit good locks to the cage.   Magnetic catches and coiled springs are not usually strong enough.   A lost hamster can sometimes be caught at night by baiting a sloping glass with food.   Once

Baited jar

inside, the hamster will not be able to climb up on the glass sides of the jar, but if plenty of bedding material as well as food has been provided, at least it will not feel the cold. It would, however, suffer stress from trying to escape such a trap if left there long, so it is very important to inspect the jar early next morning.

Temperature The preferred temperature range of golden hamsters is 20°C–24°C. Although they can thrive at much lower temperatures, in this country it is necessary to provide very generous supplies of bedding, particularly during the coldest months. They are compulsive nest-builders, and hay, white paper or medicated vegetable parchment are all suitable materials. Straw is too rough; cottonwool or knitting wool, if eaten, can cause serious internal blockage; the ink used on newsprint is poisonous. A layer of sawdust on the cage floor is important for warmth, as well as for cleanliness. Cages should be kept indoors, or at least in an insulated outbuilding. At temperatures below even 19°C weaklings, and certain strains of golden hamster, will hibernate for short periods of 2 to 6 days. A hibernating hamster, curled up into a cold, hard ball with barely perceptible breathing, is easily mistaken for dead. Sudden awakening—or prolonged hibernation—are both fatal. Pet golden hamsters do not now hibernate so commonly as in the past, for it has been shown that the tendency to hibernate is passed on genetically, and dealers are eliminating the condition by selective breeding. Should a pet be found in the state of hibernation, it is best revived slowly in the hands, or in a warm room.

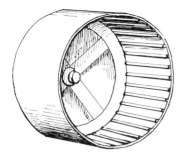

Solid
exercise
wheel

Exercise Recent research has shown that a golden hamster will run as far as 8 km a night on an exercise wheel. This is a very high level of activity for such a small rodent, and suggests that caged hamsters are particularly liable to suffer stress and lose condition as a result of forced inactivity. Although there has always been some controversy about the value of exercise wheels, at least a hamster cannot get its tail caught, and many owners find a solid wheel useful for an adult golden hamster. The wheels are not recommended, however, in the nursery or colony cages.

In summer, a hamster which is reasonably active in the afternoon, as some are, can have a play-pen in the garden. The picture below shows how one may be made from four planks and four bricks. It would not be safe

Temporary play-pen

to leave the play-pen unattended, but hamsters would benefit from a short playtime on the lawn. When outside play is impossible, it is important to allow a pet hamster a period of exercise indoors, although that too has to be supervised unless there is a very safe area available.

Cleaning The cage needs thorough cleaning once a week. It is usually thought that if the food store behind or under the sleeping nest is noticeably disturbed more often than weekly, the hamster will become distrustful, restless, and liable to bite, but it is possible to remove any decaying food from the nest daily without distressing the hamster. Other daily cleaning involves removing wilted greenstuff or fruit from the cage, changing the drinking water, and cleaning out the damp corner. This is not difficult if a small tray—perhaps a tin lid—is put under the sawdust in that particular place.

Grooming The splendid condition of a really fine hamster is the result of good breeding, sufficient exercise and correct feeding, rather than grooming. Golden hamsters are fastidious creatures, quite able to groom themselves adequately. They smell only if kept in insanitary conditions where urine has been allowed to soak into the cage, or food left to turn sour. A well-cared for pet does not smell.

Feeding In the wild, the golden hamster is presumed to be mainly herbivorous, eating dry seeds found in the desert. It would also take green food as available, and probably insects and grubs whenever possible. In captivity, a tablespoonful of cereal food a day is sufficient, and usually fed in the evening. Some hamsters prefer a dry diet; others like their cereal damped with milk into a mash. When a mash is given, the hamster should also be offered some hard food in the form of a rusk or dog biscuit. Suitable cereal foods are porridge oats, wheat, brown bread, puppy meal, unsweetened breakfast cereals, and occasionally biscuit and cake. This basic diet has to be supplemented with animal protein, nuts, fruits and vegetables in order to supply all the essentials of a nutritious diet. Packeted or pellet food also needs supplementing with fresh foods.

Fully
pouched
hamster

Animal proteins are best given in the form of hard-boiled egg, cheese, flakes of cooked fish and small pieces of cooked, lean meat. The most popular seed is that of the sunflower. Favourite fruits include grapes, raisins, and segments of apple or pear, but not citrus fruits. Favourite vegetables include raw cabbage, lettuce, Brussels sprouts, celery, spinach, chicory, watercress, parsley, tomato, carrot, and swede. Golden hamsters eat a surprising amount of greenstuff, and many wild plants, provided they are well washed, are suitable when young and fresh: dandelion, groundsel, clover and plantain, for instance. Chocolate is poisonous; nuts make a better titbit.

Young hamsters begin to feed on solid food from a very early age, and a sprinkling of puppy meal or fine oatmeal with some very finely chopped greenstuff may be scattered about the nursery cage for them from the 7th day. Pregnant females and nursing mothers need a diet with a much higher protein content than usual, and the amount of food given has to be increased gradually once the litter is born if all the pups are to be well fed. Dry Complan is a useful supplement at this time.

Pouching Food Occasionally, when a golden hamster is given its food in the evening, it will eat some of it immediately; more often, it will pouch the food, and carry it off to its store to eat later. The front paws are used as hands to pouch and to de-pouch the food, and because a

83

hamster will pouch anything, straw, sharp seeds and oat husks must not be given in case they damage the delicate lining of the pouches. The advantage of the hoarding instinct in captivity is that it is possible to leave a hamster unattended over the weekend if need be.

Drinking Water   Access to clean drinking water is essential.   This is particularly necessary for pregnant females, nursing mothers, and hamsters left for the weekend with only dry food.   Water is best given by way of a drip-feeder.   Milk, or diluted milk is also good, and if a particular hamster is reluctant to drink it, dried milk such as Marvel may be added to the food.

Handling   Unless it has been accustomed to being handled from infancy, it takes several weeks to tame a golden hamster.   Begin simply by stroking it at feeding time, and when lifting use cupped hands.   Hamsters should never be disturbed from sleep, but quickly respond to their owners if given some attention each day.   In schools, most become tractable if handled gently and allowed to get to know the children in turn.

Ailments   The most feared hamster disease is WET TAIL, characterized by a pale discharge from the anus, although the initial symptoms are lack of condition and loss of appetite.   Decline is rapid, and only hamsters which receive prompt veterinary treatment have any hope of recovery.   Because the disease spreads rapidly, the strictest hygiene with thorough disinfecting of the cage and burning of bedding is essential, especially if more than one hamster is kept, or if the cage is to be re-used.

A sudden change in the type of greenstuff given, unwashed greenstuff contaminated by chemical sprays, or a dirty cage may cause DIARRHOEA.   CONSTIPATION is likely to be due to shortage of greenstuff and water, and should be quickly cured as the diet is adjusted.   If it is suspected that the hamster has been eating cottonwool, or something similar, veterinary advice is needed.

ABSCESSES are usually the result of a wound inflicted by one hamster on another.   House the animals singly, and bathe the wound with a mild antiseptic until the abscess breaks and all the pus is cleaned away.   POUCH ABSCESSES are more serious. As there is no saliva in the hair-lined pouches to clean a wound, the smallest scratch from a sharp seed or thorn can quickly become infected. Symptoms are a swollen face, discharge from the eyes, and laboured breathing. Once the pouches become too sore to use the hamster will not feed, and decline can be rapid. Again prompt veterinary treatment is essential.

Many fly sprays are poisonous to fur-bearing animals, and should not be used in a room where animals are kept. The frequent grooming of a hamster makes it particularly susceptible to this kind of POISONING.

The hamster has no EXTERNAL PARASITES of its own, but can become infected with fleas from other domestic animals or dirty bedding. If this happens, use a proprietary flea powder on the coat and change the bedding which will harbour flea eggs and larvae.

The Mongolian Gerbil *Meriones unguiculatus*

Mongolian gerbils (the *g* is soft as in gentle) are among the most attractive of the pet rodents: alert, inquisitive, agile, and perky. Originally captured for use as laboratory animals, their charm is such that they were soon in demand as pets.

Description The dense fur of the Mongolian gerbil's back is a golden sand colour; 'ochre' is the word often used to describe its yellow tinge. The under hairs, in

Mongolian
gerbil
showing
black-tipped
tail

Albino Mongolian gerbil

fact, are slate grey, but this colour does not show through the longer guard hairs, some of which are dark along their entire length, while others are tipped with black or gold. There are much lighter patches of fur behind the ears, and around the eyes; a dark line along the tail, and a dark tuft at its tip. The fur on the underside of the body is very light: pale grey under hairs beneath white, or white-tipped guard hairs. The five claws of each limb are black. Combined body and tail length is about 18 cm.

Absence of Varieties Although albinos have appeared, and other colours are occasionally reported, breeders have so far failed to produce uniform colour varieties such as those recognized for hamsters, guinea-pigs, mice, and rats. Inbreeding is the method most likely to produce colour mutations, but it is always hazardous; deformed offspring are common, and in the case of the Mongolian gerbil it was abandoned in the early breeding programmes in favour of breeding by random selection, whereby the risk of deformity—but also the incidence of colour mutation—is far less.

Origins Breeding in captivity began in 1935 when a

Japanese scientist, Kasuga, captured 20 pairs from the Amur River basin on the Sino-Soviet border (map below). From Japan, in 1954, 4 breeding pairs were sent to America from where, in turn, the first 12 breeding pairs reached Britain in 1964, followed soon afterwards by another 12 pairs. From these, practically all our subsequent laboratory and pet Mongolian gerbils derive.

'Yellow Rats'   The Abbé Armand David was the first European to record seeing the Mongolian gerbil. In his Diary they are referred to as 'yellow rats' (*Rats jaunes*). When that great French naturalist saw his 'yellow rats' in 1866, he was somewhere near the Inner Mongolian/Chinese border, in a region covered by the so-called 'yellow' loess soils of Asia, and drained by the great 'Yellow' River. It is not then surprising that he used the same adjective to describe the similar, and presumably protective colouring of one of that region's most abundant rodents. Père David sent specimens of these 'yellow rats' to Paris where, in 1867, they were classified by Henri Milne-Edwards as *Meriones unguiculatus*.

Sketch map of Far East

Natural Habitat   We lack information on how far this gerbil penetrates into the Gobi, but although the natural habitat is not accurately defined, it certainly includes much of the desert, and semi-desert of the high plateau of Mongolia, Inner Mongolia, Manchuria, and the Shansi and Kansu provinces of China (p. 87).   These mid-latitude deserts of interior Asia are, in one way, an even harsher habitat than the hot deserts.   In addition to aridity, and to a marked diurnal range of temperature during the hot season, animals here have to withstand the extreme seasonal variation associated with continental interiors: high daytime temperatures in summer, sometimes in excess of 43°C, and sub-zero temperatures in winter.

Adaptations to Natural Habitat   Mongolian gerbils are well adapted to survive aridity, and use water very economically.   There is a general lack of sweat glands to prevent evaporation of water from the body surface, and only a few drops of highly concentrated urine are excreted. It is even possible for these gerbils to live through periods of severe drought without drinking at all, eating only dry seeds.   Their survival is assured by the retention of so-called 'metabolic water', produced when starch in their diet is ultimately broken into carbon dioxide and water.

   Fur is an effective insulator against both cold and heat, but probably the Mongolian gerbil's most important protection from extremes of temperature is achieved by burrowing.   In a network of tunnels from 0·5–1·5 m underground, the gerbil attains a tolerable and almost equable temperature in its living quarters, but these burrows are probably more important in giving protection from cold rather than heat.

   It is the harsh, continental winter that must be the worst time of year for them, and we know they are less active in the cold season, remaining underground for long periods in the severest weather, relying on metabolic water, and on stored supplies of seed.   Gerbils in cultivated areas are not likely to suffer from lack of food, for the loess is fertile when irrigated, supporting crops of both wheat and millet.   It is in uncultivated areas that the Mongolian gerbils need to travel long distances to search diligently for dry, windblown seeds, and when there is rain, and the desert flowers briefly, they doubtless take green plants for their water content.

Longer hind limbs help the gerbils move fast when chased by predators, and the tail is also a help. It acts as a balance, allowing the Mongolian gerbil to leap and turn to escape with its life. The black tuft at the tip of the tail is in itself a safety device; if held by the tuft, the gerbil will slip it, leaving the vertebrae exposed, but escaping once with its life that way. The tuft never grows again. It is likely too, that away from the home burrow, gerbils dig emergency bolt holes where they may seek sanctuary in a featureless environment when hunted, or alerted to danger by their keen sense of hearing. But in spite of all these evasions, the gerbil must often fall prey to its hunters. Small rodents are essential in the ecology of the desert as food for the flesh-eating birds of prey, and for the carnivorous mammals.

Life Cycle in Captivity　Birth takes place, most often at night, after a gestation period of 24 days. A litter of about 6 young may be expected, although litters of 1–15 have been claimed. Mongolian gerbils are born hairless, with a conspicuous red skin colour. Birth weight is ordinarily 2·5–3·5 g, increasing to 11–18 g by the age of 4 weeks. Within a few hours of birth the young will move around and cry a little. During the 1st week of life the ears open (5th day) and hair begins to grow (6th day). In the 2nd week the incisor teeth erupt (13th day). By the end of the 3rd week the eyelids are open (16th–20th day). The mother will suckle her young for at least 3 weeks, and sometimes for as long as 4 weeks. After weaning, development continues to be rapid and the young females reach sexual maturity some time between 9 and 12 weeks of age, and the males normally at 10 weeks. It is possible for the female's reproductive life to extend to 20 months, the male's to 22 months, and possibly beyond. The life span in captivity is about 3 years.

Male and Female　In spite of inevitable variation in the size of individuals, the male may be expected to be heavier than the female. Average adult weights are 100 g for females; 117 g for males. With age, both tend to put on excess weight.

The male's tapered body is accentuated by a distinct fringe of long hair which shows at the butt of the tail. The apparent lack of nipples is of no help in determining the sex of an adult gerbil. Only the females bear them, but even during lactation these gerbils retain a thick coat on the underside of the body, and the nipples are hard to detect, although they do show on juveniles before the hair grows. In sexing young Mongolian gerbils it is easiest to note the distance between the anus and genital opening which is greater in the male. The dark-coloured scrotal sac already shows on males by the time they are weaned.

Breeding Experience has shown that Mongolian gerbils are best kept in pairs, and housed together for life. They are capable of raising perhaps 9 litters, and afterwards, in old age, show a touching affection for each other, indulging in much social grooming. Both are good parents, and at the time of the birth and rearing of the young it is lack of privacy and undue human interference that is unwise, and can lead to a nervous mother killing her young. The presence of the male is in no way detrimental. He takes an active part in building a nest from the paper he shreds, and in keeping the young warm. If young gerbils have to be handled it is prudent first to remove the mother so that she at least does not witness the interference, but the danger of her turning against her young, after their disturbance, is much less than with mice, rats, and hamsters.

Pairing When buying a pair of Mongolian gerbils for breeding, select a healthy unrelated pair who have already accepted each other and have been housed together by the breeder. The best time to buy is when they are between 6 and 8 weeks old, and certainly before they reach 10 weeks. It is always difficult, and sometimes impossible to pair-up adult gerbils, even for mating. Attempts to do so often lead to fighting and to the infliction of fatal wounds.

After the death of its mate, a particularly fine specimen still in good breeding condition might be successfully introduced to a new mate in a quiet room on neutral ground, and not in a cage at all, and then the pair re-housed together in an environment new to both. If fighting were going to happen it would begin quickly, and the gerbils should be separated immediately. Possibly, though, on another occasion the introduction might succeed, and that chance would warrant a second attempt.

Oestrous Cycle   The female comes into season approximately every 6 days throughout the year, and remains on heat for at least 5 hours at a time.   Mating is most often observed in the early evening, one's attention being attracted by the male's drumming of his hind feet as at other times of high excitement.   When the young are born the placenta (afterbirth), together with any stillborn offspring, will be consumed by the mother.   She should not be disturbed.   Many females come into season again immediately after giving birth, but if mating does not occur very soon after the birth of a litter, the female will not usually accept the male until she has weaned her young. Breeders habitually remove a litter from the mother on the 22nd day, for they expect the birth of another litter on the 24th.   Fortunately, pet owners do not often have to contend with breeding cycles as intensive as this, and unless the mother is again pregnant it is best to leave a litter with her for 28 days, unless she tires of the youngsters earlier, and pushes them from her.   A casual observer may not be able to detect a gerbil's pregnant condition, but many pet owners will notice the increase in body size. The weight gain will be about 10–30 g, and it is as well to withhold some sunflower seeds from a pregnant gerbil, if she shows a particular fondness for them, to prevent her becoming overweight.

Colony Rearing   There is sufficient evidence to show that in their natural environment Mongolian gerbils live in underground colonies, presumably in extended family groups.   In captivity, colony rearing consistently fails because of serious fighting.   When first taken from their parents, the young may be raised as a colony up to the age of 6–8 weeks, but by then they will have begun to squabble, to fight, and to pair-up.   One assumes that in the wild, large family groups extending over several generations, while sharing the same network of burrows, have carefully bounded territories within it.   These boundaries are most likely marked by the secretions of the

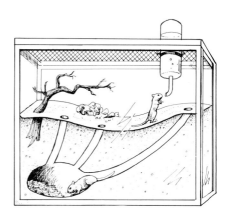

Gerbilarium

abdominal sebaceous gland that is present in both sexes. If this should be so, the limitations of space available for colony rearing in captivity—quite apart from the effects of other stress factors—would alone lead to encroachment of territorial boundaries, and to the consequent fighting that makes this method of rearing impossible for mature gerbils.

Housing in a Gerbilarium Nevertheless, gerbils are social animals, and if you are not keeping a breeding pair, keep two females. They will live peaceably together for life, if they too are selected before the age of maturity, and introduced to their living quarters together. The recommended way to house Mongolian gerbils is in a gerbilarium where a pair can burrow, to some extent, as in the wild. The burrowing medium is a mixture of slightly damp peat, potting compost and straw. It must be rammed down hard to make a firm mound, and may be covered with grass turfs. In practice it will be found best to slope the mound, leaving the animals plenty of headroom to move about above ground level when they wish. The advantage of using this mixture is that the tunnels, once excavated, will hold their shape; other burrowing mediums often fail because the tunnels collapse.

There is no risk that once housed in a gerbilarium, the

gerbils will be lost to sight. In captivity most display a genuinely diurnal habit, being active throughout the whole 24 hours of the day, alternating periods of intense activity with periods of sleep. One great advantage of a gerbilarium is that it affords the gerbils the luxury of privacy in the nest compartment for their rest periods which are vital to their health (cf. Ailments).

Temperature The ideal temperature for gerbils is in the range 20–24°C, so a glass gerbilarium should be shaded from direct sunlight that would cause it to overheat; and in order that the animals may be comfortable when the surrounding temperature falls, plenty of clean paper should be given for them to shred into bedding material. The poisonous quality of printer's ink makes newspaper unsuitable. With very young gerbils the question of temperature is vital. Statistics suggest that not more than 70% of live-born gerbils are successfully weaned, death by chilling being a contributory factor. One problem is that very young gerbils are unable to shred paper. When they are first removed from parental care, ready prepared bedding needs to be provided.

Housing in Cages Although much less satisfactory, the more common way to house gerbils is in a cage. A well designed home-made cage may be superior to a manufactured one, but if made of wood should be strong enough to withstand constant gnawing. The gerbils can be given the benefit of extra floor space in the cage by the addition of ramps and galleries that allow for more movement within a confined space. They need a really deep litter through which they will bulldoze their way once proper burrowing is denied them. Gerbils often ignore a sleeping compartment, arranging their bedding in the open cage to suit the temperature, but it is as well to provide a nest box, partly for extra protection in cold weather, and partly as a place of refuge.

Exercise Gerbils so often use their few possessions and toys in novel and ingenious ways that they deserve a well thought out home that will be a stimulating environment for them. Simple possessions such as a wooden toy, or a

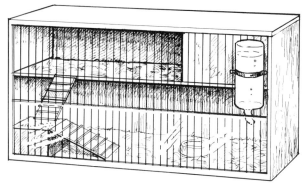

Two-storey cage suitable for gerbils

cardboard tube, give them hours of play. As rodents they need wood to wear down their incisors, but they will climb and balance on a branch as well as gnaw on it. Some owners provide an exercise wheel, but it should not be a fixture, as gerbils are likely to overtax their strength by using it excessively. Gerbils may also be exercised out of the cage, and in good weather, out of doors if they have a wooden or small-mesh play-pen for safety. Cats and dogs are not to be trusted. If they will hunt—given the opportunity—other small rodents, they are just as capable of taking gerbils.

Cleaning Every day a gerbilarium will need attention: stale vegetable matter should be removed, the water bottle re-filled, and so on, but major cleaning need not be frequent. Gerbils do not usually hoard food in captivity, so the problem of having to disturb a gerbilarium periodically to get at stores of decaying food does not arise. As gerbils excrete such small amounts of urine, a gerbilarium with a deep layer of highly absorbent peat can be left for some weeks—at least half a term—before being cleaned and replenished with a fresh peat and straw mixture. In practice it will usually be found necessary to clean a cage rather more frequently—about once every 3 weeks. Unnecessary cleaning will be avoided if care is taken to see that the gerbils do not pile up litter beneath their water bottle to make it leak.

Grooming As far as grooming is concerned, Mongolian gerbils need no help from their owner. A pair of gerbils will keep themselves, and each other, scrupulously clean and, except sometimes in old age, do not smell when properly housed in clean, dry accommodation.

Feeding Mongolian gerbils living in the wild are classed as herbivores. It is thought their diet must be made up almost entirely of seeds, grains, and green plants. Although we lack any evidence, it may be that they take a little animal food in the form of inverte-brates. Certainly it is not unknown for a pet gerbil, allowed the freedom of a room for exercise, to take a passing spider. Gerbils are able to digest animal protein, and the addition of a little to their diet has been found beneficial in captivity, particularly to females during pregnancy and lactation, and to newly weaned gerbils with a high growth rate to maintain. It is most easily supplied in the form of dried skimmed milk, hard-boiled egg, or a few pieces of dog or cat food. But, in the main, gerbils may be fed a dry, herbivorous diet, that is a mixture of canary seed, sunflower seed, wheat, oats, maize, and barley. This basic diet should be fed at the rate of one tablespoonful a day for each adult, and adjusted according to how readily the gerbils take the titbits offered at other times. Favourite titbits are melon seeds, raisins, potato-crisps, and unsalted peanuts. The basic diet then needs to be improved by the inclusion of a little well-washed fruit or vegetable.

When feeding pellet food, a high protein rodent one is suitable, supplemented with a little fruit or vegetable. Pellet food has the disadvantage of monotony, whereas in practice it will be found that gerbils, like most other rodents, enjoy a wide variety of foods, and relish the occasional new taste. Just for this reason pellet food is perhaps best fed only when the numbers of animals kept necessitate its use for reasons of convenience.

The only food Mongolian gerbils are likely to overeat is sunflower seed. When measuring the supply of food to

Cupping gerbil in hands

be left with gerbils for weekend feeding in schools, avoid leaving a high percentage of this seed, for the gerbils are likely to eat it immediately, leaving themselves short of food for the rest of the period.

Drinking Water Water is best given by means of a drip-feed bottle suspended from the lid of a gerbilarium, or clipped to the side of a cage. Pet owners should not be misled, by the apparently unchanging water level, into thinking their animals are not drinking. Careful measuring has shown that, given the freedom to drink at will, even gerbils with access to plenty of fresh fruit and vegetables will still take about 4 ml of water a day. Bearing in mind that a standard teaspoon has a 5 ml capacity, a 4 ml drop in level will not show in most bottles.

Old gerbils, and also females after giving birth and throughout lactation drink thirstily, sending up streams of air bubbles into the water bottle.

Taming and Handling It is as well to accustom Mongolian gerbils to being handled from the time they are weaned, if they are to be tame when adult. At first there is no need to pick them up, for their natural curiosity is such that they will explore a hand—or anything else—put near them. Once the creature's confidence has been gained in this way, the Mongolian gerbil will climb onto a proffered hand and may then be cupped in it and lifted gently. Avoid overhandling, restraining by the tail, chasing a gerbil around its cage, and dropping a hand down onto its back. Perhaps because this movement resembles the attack of a bird of prey, it invariably frightens a gerbil.

Ailments Quite often the weakest gerbil(s) of a litter may decline if taken from the mother as early as the 22nd day. The youngster(s) should be returned to the nest for a few days. They are usually accepted, even if the mother, meanwhile, has given birth to another litter. Several attempts at FOSTERING UNWEANED GERBILS have also been successful.

SEIZURES that completely prostrate a Mongolian gerbil are alarmingly common, and will become more so while affected animals are allowed to breed. The seizures are of an epileptic nature and very likely passed on genetically. There is some suggestion that in nature they may be used as a ruse to feign death in the face of a predator, but in captivity they are commonly induced by overhandling, alarm, and disturbance of the sleep pattern, and frequently lead to actual death.

In common with other rodents which have to gnaw on metal and wiremesh in captivity, the Mongolian gerbil can also suffer from SORENESS and SKIN LACERATIONS. Bathing with a mild antiseptic helps, but the condition recurs unless the creature is more suitably housed, and given wood to gnaw.

SORE EYES are most likely caused by the irritation of smoke, dust, or unsuitable litter or bedding material to which the gerbil shows an allergic reaction. An attempt should be made to identify and remove the irritant, meanwhile bathing the eyes with an eye lotion. A case that does not respond speedily to home treatment should be referred to a veterinary surgeon.

WOUNDS are frequently inflicted by one gerbil upon another unless the precautions for introducing gerbils, outlined above, are followed carefully. If the wounds are minor, it is sufficient to bathe them with a mild antiseptic, but deep wounds need veterinary treatment, and the animal may have to be destroyed.

The Rat and the Mouse

Rattus norvegicus and *Mus musculus*

Origins These two closely related and excellent pets have been bred from most unpopular ancestors. Tame rats were raised, by selective breeding, for laboratory work at the beginning of the century. They are mutant forms of the Norwegian, or brown rat, *Rattus norvegicus* (Berkenhout), which has spread throughout the world and is usually found in association with human habitation. The wild form has always been a major pest, feared as a carrier of disease as well as for its depredations. The tame rat, by contrast, is as useful as the wild rat is destructive. It is second only to the mouse as the most popular laboratory mammal, and so attractive that it is now an established pet, particularly well adapted to cage-life.

Fancy mice have been raised from another commensal animal, the house mouse, *Mus musculus*, Linné, which has also spread throughout the world, wherever man is, or grows his food, or stores it. Like the rat, the house mouse is a destructive pest, but was bred in captivity at the end of the 19th century in an effort to find a small, prolific, laboratory animal. The project was so successful that the mouse is now the most commonly used laboratory mammal, and so attractive in its mutant forms that it has become a very popular pet, in spite of its origins.

Broken-marked mouse and Dutch mouse

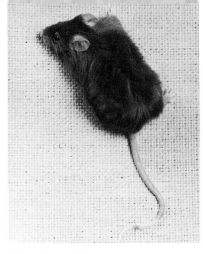

Long-haired
mouse

Varieties After 75 years or more selective breeding, more than 40 varieties of fancy mice are recognized by the National Mouse Club. They are catalogued for show purposes under four headings:

Self Varieties The self colours include white, black, blue, chocolate, red, fawn, champagne, silver, dove, and cream.

Tan Varieties These are mice of standard colour with a tan underside. They are known, according to the top colour, as black and tan, blue and tan, etc., and are most often bred with black, blue, champagne, silver, and dove top colours.

Marked Varieties These are the patterned mice. The varieties most often seen are broken-marked mice which have as many coloured spots or patches as possible distributed unevenly over a white ground; and Dutch mice (p. 98) which are marked as the Dutch rabbit (p. 47).

Any Other Varieties This is a large group and includes agoutis and cinnamons; chinchillas; sables; silver greys, browns, and fawns; pearls; Argentés; seal point Siamese; and chocolate Himalayan. Two others are mice of different coat types: the long-haired, and the astrex, which is the curly-coated mouse.

99

By contrast there are only a few rat varieties, and there is not nearly such a thriving rat fancy in this country at present. Laboratory breeding gave rise to two **albino strains** of comparable size to the brown rat, and to a smaller, distinctive, bicoloured strain. This is known as the **hooded rat**: a white rat 'hooded' by a coloured head and shoulders, the colour extending down the line of the spine. The originals had a black hood, but chocolate, cinnamon, fawn, cream, and agouti are now seen.

Disconcertingly, however, hooded rats are not always hooded. This same strain is bred in a range of self colours including black, white, chocolate, cinnamon, fawn, cream, and agouti. There is also a black with white feet known as the black Irish.

Description Both rats and mice are rodents of the sub-family Murinae. Characteristics include the pointed muzzle, the split upper lip revealing the rootless (i.e. permanently growing) incisors, and the ability to use the forepaws in feeding. The long tail, accounting for about half the animal's total length, is used as a balance in climbing. The phenomenal climbing ability of mice is well known, but the rat too is a competent climber, although its wild ancestors are more often thought of as subterranean animals.

In captivity, both tend to follow the nocturnal or crepuscular habit of the wild forms. It is noticeable that mice become inhibited once they become aware of the presence of man; rats tolerate human presence much better. For them, human companionship may compensate for loss of their own kind. Rats do tolerate being caged singly, especially if they are given lots of attention, but it is preferable to keep two. Males are more amiable than females, and will live peaceably together if neither has seen a female. With their own kind, male mice can be very aggressive, and must be housed singly from the age of 5 weeks. Female mice are best kept together in small colonies.

Male and Female Female rats and mice are known as does; males are known as bucks. Both can be sexed from birth by comparison of the distance between the anus and genital opening, the distance being much greater in the

Self-coloured mouse

male. Adult females show two rows of teats. The teat spots show on newborn specimens, but are soon hidden by fur.

Adult weights vary with the strain, as well as in-dividually. This variation is particularly marked in rats. Hooded rats may attain adult weights of 350 g for a male; 250 g for a female. The larger albino strains may be expected to reach 500 g for a male; 350 g for a female. Adult weights for mice are in the range 35 g for females; 50 g for males.

Breeding In order to control their numbers, it is advisable to put the sexes together only for breeding. Mice breed better in company. It is usual to establish a harem of 1 buck mouse with 2 or 3 does, but a single pair of rats will breed successfully.

The does come into season every 4–5 days throughout the year, and immediately after giving birth. Always introduce the doe to the buck's accommodation, and once it is clear she is in kindle, move the buck to another cage. If left with the doe there is a danger of his remating her after the young are born, or of harming the young. All pregnant females of a mouse harem may be left together. They will almost certainly share the same nest, and suckle the babies indiscriminately. Pregnant rats, however, will not tolerate the presence of another female, although it is not, in their case, essential to remove the buck.

The gestation period is about 3 weeks; usually 20–21 days for mice and 22 days for rats. A few days before the litter is due the cage should be cleaned thoroughly, and plenty of nesting materials supplied, including good quality meadow hay, and paper towels or tissues. The litters are large, commonly 6–8 mice, and 6–12 rats. Litters of 20 are not unknown. If it is possible, deformed or still-born young should be removed unobtrusively. On no account should live-born young be touched, unless it is to reduce the size of the litter to a manageable number. Any interference before the young are about 10 days old can cause the mother to resort to cannibalism.

Birth weights vary with the litter size; the smaller the litter, the larger the individual young. They are born very immature—hairless, with eyes and ears closed. Hair begins to grow within the first week (3rd and 4th day); and ears and eyes are open by the end of the second week (10th and 13th days). During the third week the young will be moving around, taking a little solid food such as bread and milk or fine oatmeal. Weaning begins at 3 weeks, and the mother may be removed as soon as it is certain that the young are self-sufficient.

Young mice should be segregated, according to sex, by the 35th day to prevent unwanted pairings; rats mature later, and may safely be left together until the 45th day. Breeding is best delayed until buck mice are 10 weeks old, and does are 12 weeks old. Individuals vary enormously. Many mice are past their active breeding life by the age of 9 months; others breed successfully to the age of 15 months or so. Rats are at their best between the ages of 3 and 10 months; they are rarely capable of reproducing after the age of 18 months. Life span in captivity is about 3 years for rats; 2 years for mice.

Housing Both rats and mice are such inveterate gnawers—and mice such determined escapers—that soft-wood cages would never contain them. Suitable cage-

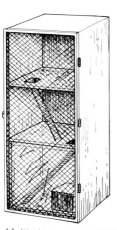

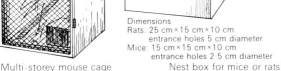

Multi-storey mouse cage

Dimensions
Rats: 25 cm × 15 cm × 10 cm
          entrance holes 5 cm diameter
Mice: 15 cm × 15 cm × 10 cm
          entrance holes 2·5 cm diameter

Nest box for mice or rats

building materials include hardwood, moulded plastics, metal, glass, weldmesh, and wiremesh—the mesh being less than 6 mm to exclude house mice.

The multi-storey cage is particularly suitable for mice, allowing them to indulge their talent for climbing. The minimum recommended base-size is 45 cm × 30 cm, with 25 cm headroom in each storey. Avoid cages in which the animals could fall from top to bottom. The more traditional single-storey cage with raised gallery should have a base-size of 60 cm × 30 cm. Mice need ropes and ladders for climbing, a wheel for exercise, a small bark-covered log or wooden cotton reel for gnawing, and a nest box. They are subject to great stress if they cannot find a dark refuge within their own cage.

Rats can also be kept in a single-storey cage with raised gallery, recommended minimum size being 75 cm × 30 cm × 30 cm high for a pair of rats. A nest box is not essential for rats, who will always construct a nest from the available bedding, but it is recommended. Rats need a gnawing block, and provision for tunnelling as well as for climbing. Lengths of piping serve as substitute burrows.

Temperature The preferred temperature range for rats and mice is 20°C–25°C, and although both can tolerate lower temperatures, cages must be kept indoors in this country. The animals' own judicious use of the available bedding material will enable them to control their temperature to some extent. Always supply plenty of bedding, including good quality meadow hay. This should be placed on top of a 2·5 cm layer of wood shavings, peat moss litter, or proprietary floor-litter, spread throughout the cage.

Exercise Rats and mice should not be allowed together, nor with any other animal, but both are highly active, inquisitive creatures, needing daily exercise outside their own cage. Rats are intelligent enough to enjoy the challenge of finding their way to a goal—some prized titbit—through a maze constructed for them by an understanding owner. They also seem to appreciate freedom to explore their own room. Within the cage a solid exercise wheel (p. 81) will give some much-needed activity, but avoid open wheels in which the tail could get caught. Wheels are particularly popular with mice.

Cleaning and Grooming Rats and female mice do not smell if kept in hygienic conditions. There will be some smell if buck mice are kept. Floor-litter will need renewing at least twice a week; cages will need scrubbing at least once a week. After scrubbing, rinse thoroughly, and allow to dry before renewing the floor-litter. Keep a spare cage to house the animals while their own is being cleaned. They will be less disturbed if a certain amount of used bedding—which will be impregnated with their own scent—is returned to the cage. Similarly, avoid disturbing 'hoarder' rats unduly. 'Hoarders' are rats which collect treasures—such as the contents of the wastepaper-basket—for their nest. Allow them to keep all but the precious or perishable part of their spoil.

Both rats and mice are fastidious about their personal hygiene, and will groom themselves adequately, without help from their owners.

Food and Water Rats and mice are omnivorous in captivity. It is usual to feed two meals a day, served in

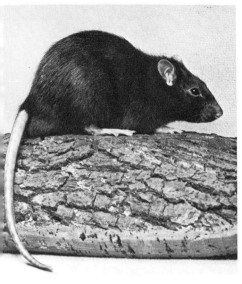

Black
Irish rat

dishes that can be scalded each time they are used. The morning meal is a dry mixture of whole or rolled oats, with some bird seed. For preference use a seed mixture for canaries, because it will contain high-fat seed such as rape, linseed, niger, or hemp. A mash of wholemeal bread and milk, squeezed to a crumbly consistency, is not fed until the evening, when these nocturnal animals are at their most active.

Raw fruit and vegetable should be given daily. Favourites include carrot, swede, celery, and segments of apple. Green food should be fed only in moderation—dandelion, watercress, and raw spinach being considered particularly good. Hay is important to both animals. In addition, suspend a salt and mineral lick in the cage permanently.

Occasionally the basic diet of grains and cereals may be supplemented with a little animal protein in the form of hard-boiled egg, or pieces of dried dog or cat meat. Pregnant does are particularly in need of an enriched diet, and should be fed bread and milk *ad lib*.

Drinking water is best provided in drip-feed containers.

Methods of handling mice and rats

Open dishes quickly become spilt and fouled.   Rats drink
more than mice, and in a cage housing several, two drip-
feed bottles may be necessary to meet their needs.   Rats
and mice fed on pellet food, which is not normally a
necessary expedient for pet animals, will drink very much
more water, even if the pellets are supplemented with
plenty of vegetable matter.

Handling   In spite of their undoubted agility, rats and
mice are easily handled, rats in particular being docile and
gentle with man.   Both become tractable if handled frequently from the time they emerge from the nest.   They
are most easily tamed by feeding from the hand, a practice
it is wise to continue permanently to maintain their trust.
When lifting, rats or mice can be scooped up in both
hands.   Mice may be caught by the tail and transferred
immediately to the other hand.   Retain a hold on the tail
to forestall a sudden leap.   It is accepted practice to scoop
up a docile rat one-handed, or to place one hand over the
back with the head held between thumb and forefinger.

Close the remaining fingers beneath the belly, or support the weight with the other hand.

Ailments In common with rabbits, gerbils, guinea-pigs, and hamsters, these small rodents have very poor recuperative powers.   The accent, therefore, must be on the prevention of illness by very high standards of care. Immediately a slight loss of condition is noticed, the animal should be isolated from all others, and veterinary advice sought.

Mice and rats are prone to respiratory tract infections such as BRONCHITIS and PNEUMONIA.   Avoid sudden changes in temperature, draughts and dampness which may lead to these conditions.   Another common complaint is an ALLERGIC REACTION to dusty hay or saw-dust.   Use only best quality hay, and softwood shavings rather than sawdust as a floor-litter.

INTESTINAL COMPLAINTS can often be avoided if good quality food is fed, and always in clean utensils.   Never leave a mash in the cage long enough to turn sour, and replace vegetable matter daily.

EXTERNAL PARASITES are not often seen on rats and mice, but if they occur they have probably been brought in on contaminated bedding.   Use a cat flea powder, and after scrubbing the cage replace the floor-litter and bedding.

If a rat or mouse stops feeding, check the length of the INCISOR TEETH which may have become overgrown.   A veterinary surgeon will trim the teeth, but afterwards make sure the animal can always reach a gnawing block.

CONGENITAL ABNORMALITIES include deformed young and a middle ear disease that affects the balance causing the animal to hold its head awry and to turn in circles. Consult a veterinary surgeon about humane destruction of affected animals.

UNGULATES

The Pony *Equus caballus*

The pony is rivalled only by the dog as the animal so many children aspire to own. Undeniably it makes a very fine companion. The sympathy between child and mount that makes riding such a special pleasure is experienced by many, and keeps the demand for ponies constant. However, the pony is by nature a grazing animal, and the modern practice of keeping it in urban areas with little or no pasture is a very worrying trend.

Varieties There are over a hundred breeds and types of horse and pony in the world, of which more than half are ponies. The great majority of children's ponies in Britain are descended from the nine native mountain and moorland breeds: the Dartmoor, Exmoor, New Forest, Welsh, Fell, Dales, Highland, Shetland, and Connemara. Over the years Arabs and Thoroughbreds have been crossed with these breeds to produce quiet and mannered ponies of very good quality.

   The definition of the word pony is inexact, but in addition to the true pony breeds a small horse of practically any breed or type, standing not more than $14\frac{1}{2}$ hands* (approx. 147 cm), is known as a pony. There are two notable exceptions. The Arab is always referred to as a horse, no matter how small; and the polo pony is always called a pony, no matter how high. Another exception is the high-stepping Hackney which is a pony only up to the height of 14 hands (approx. 142 cm). Hackneys exceeding this height are horses.

---

* The height of a pony is measured at the highest point of the withers. The traditional unit of measurement, used for centuries before the use of the measuring stick, was the width of a man's hand. The term continues right up to the present, and now denotes a measurement of 10·1 cm.

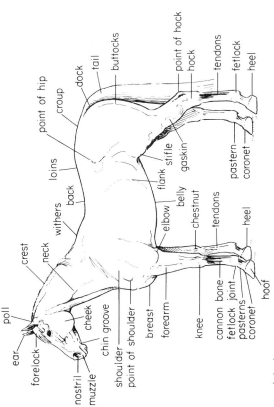

Points of the horse

109

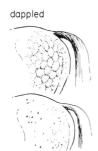

dappled

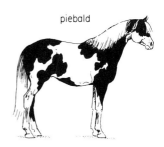

piebald

flea-bitten

Broken coloured ponies

Colours Ponies are said to be whole-coloured or broken-coloured. The recognized *whole colours* include black, brown, bay, chestnut, dun, palomino, and grey. There are many shades of some of these colours: grey, for instance, includes most white horses. By tradition, white horses are nearly always referred to as grey unless they are albino or Arab. Whatever the shade, all the body hair of a whole-coloured pony is uniform, although there may be white face markings, and the mane, tail, and points (i.e. the lower legs: foot to knee) may be a different colour. The bay always has a black mane and tail, with black points; the chestnut has a mane and tail matching, or paler than, the body colour; the palomino has a white mane and tail.

Broken Colours The main colour of roan ponies is interspersed—and so modified—by the presence of white hair. This white hair tends to lighten the overall colour, and when mixed with black produces the blue roan; when mixed with bay produces the red roan; when mixed with chestnut produces the strawberry roan. Broken-coloured ponies include piebalds with large, well-defined white patches on a black body; and skewbalds with similar patches on a body of any other colour. There are also dappled and flea-bitten greys, spotted with darker hair: dappled referring to the larger size of spot; flea-bitten to the smaller.

Description   The horse and pony, *Equus caballus*, Linné, are ungulates or hoofed animals, very closely related to the onagers, asses, and zebras, all of which, being equines, are capable of interbreeding to produce mules, and which are thought to share a common ancestor. Originally prairie or grassland animals, horses have the long face that accommodates the grinding teeth of the grazer, and the long legs that provide the defence of speed against predators.   The single toe (see paragraph below) is stronger than several toes, and horses bred for racing achieve speeds of 56 km/h.   At speed, and when jumping, there is a tremendous strain on the legs, and to absorb the shock the fetlocks flex on impact, stretching the tendons above and below.

Evolution   The horse has not always been a long-faced, fleet, grassland animal, as we know from fossil evidence. The so-called dawn horse, *Eohippus*, existed in North America 70,000,000 years ago, and was a small, short-faced animal, about the size of a wolf, with 4 toes on each foot.   It browsed on swampland vegetation.   As the swamps were gradually replaced by grassland and forest, there was a parallel evolution of *Eohippus*.   The legs lengthened and the toes contracted on the firmer ground to give an animal the size of a small pony with 3 toes. Further evolution produced, by the beginning of the Pleistocene period, about 1,000,000 years ago, the single-toed *Equus*, which is the recognizable ancestral type of our modern horses, onagers, asses, and zebras.

Although Pleistocene horses suffered extinction in the Western World, herds must have migrated to the East across still existent land-bridges before the Ice Age, and so escaped total extinction.   The domestication of the horse, therefore, took place in the Old World, and it was not returned to the Americas until the Spanish reintroduced it as a captive animal in the 16th century.   Inevitably some escaped, but thrived so well as feral animals in their ancestral home, that herds of these Mustang—as they became known—were common in North America until recent times.

Domestication   The earliest of the true horses in the Old World colonized the steppe lands of Europe and Asia

A modern partnership between horse and rider

and differentiated, according to environment, into several geographical races, from which man has created the modern breeds. Two of these races survived in the wild state into recent times: the Tarpan into the 19th century; the Przewalski into the 20th century.

The credit for first domesticating the horse seems to belong to Neolithic peoples of the Eurasian steppes, who probably domesticated the Tarpan between 3000 and 2500 BC. Cave drawings, as well as other archaeological finds, suggest that the domesticated horse was at first used for meat, milk, and hides; later for haulage; and later still—in the Bronze Age—for riding.

The horse added much to man's stature and mobility, and greatly enhanced his reputation as hunter, herdsman, and warrior. It may perhaps seem surprising that such a prestigious animal was the last of the three equines successfully domesticated, for it is certain that both onager and ass were domesticated earlier. The reason almost certainly has to do with the fact that the horse resists domination and has to be 'broken in' before serving man with its great gifts of speed, strength, and stamina.

Life Cycle in Captivity   The young are born after a gestation of about 11 months.   Normally a single foal is born, but sometimes twins.   A single birth is preferred. With twins, each animal is approximately half the weight of a single-born foal, and is difficult to rear.   A male foal is known as a colt; a female as a filly.

The young are born with eyes open.   They are fully covered with a coat which usually differs both in colour and texture from the adult coat that grows at about 6 months.   They are cleaned by the dam and are usually able to stand within 2 hours of birth, and to follow her around within about 5 hours.   They take some solid food within a few weeks, but are not weaned before 5–6 months.

The fillies are capable of breeding at 2 years, but are rarely used before 3 years.   Their breeding span usually extends into old age.   Colts are also capable of breeding at 2 years, but do not reach peak condition before 4 years.

The life span in captivity is most often in the range 20–25 years, but ages of 50 + years have been recorded.   In captivity many horses and ponies do not live their full span, their lives being terminated in old age on veterinary advice.

Male and Female   An uncastrated adult male of 4 years and over is called a stallion; the castrated male is a gelding; the female is a mare.   The males are slightly larger and thicker than the mares, and the stallion—but not the gelding—tends to develop with age a muscular, arched neck.   The stallions are also more temperamental.

The mares have 2 teats and only 36 teeth instead of 40, as in the males.   They normally lack the tushes—the canine teeth—that the males grow at 5 years.   The size of ponies varies with breed, the smallest adults standing no more than 9 hands (approx. 92 cm) and weighing just 76 kg.

Breeding   In the main, bloodstock breeding in Britain takes place on stud farms (i.e. breeding establishments), where the mares and foals are in the charge of experienced stud men. The novice is sometimes attracted to the notion of breeding a pony at home, but the idea should be resisted unless there are very good facilities, and unless the family has impressive knowledge, and some experience of breeding ponies—or can employ a good stud man. It is out of the question if only one or two small fields are available for the mare and foal. Since a good stallion can serve up to 100 mares a year, relatively few are needed for stud work. This, and the fact that the greater docility of the gelding is normally preferred, has led to the practice of castrating most yearling colts apart from race horses.

Oestrous Cycle The mares are in season for 3–7 days every 3 weeks during the summer, and 9 days after foaling. Most owners like to have their mare served (i.e. mated) in spring or early summer, so that she foals the following May or June, when there is plenty of good grass for her, and warmth for the foal.

Field or Paddock A pony needs at least 1 hectare of grass, a companion, and good shelter—preferably a loose-box as well as a field shelter. The field or paddock can be divided into approximately $\frac{1}{2}$-hectare plots so that these may be used in rotation. Each plot will, of course, need a separate water-supply and field shelter, but the extra expense is justified only if several ponies are being kept together.

Obviously, grass needs managing like any other crop, and should be treated annually with a compound fertilizer containing nitrogen, phosphate, and potash. In addition, an application of lime every 5 years will help to keep the grass sweet. Alternatively, slower-acting fertilizers such as basic slag are available. These need to be used every 3–4 years. Check that the fertilizer has been well washed in by rain before turning a pony onto the treated area. If weedkillers have to be used, keep the ponies off the land for at least a fortnight.

A field or paddock used for ponies needs a good land drainage system, but this is wasted unless maintained properly. The outlets must be kept clear, and the ditches dug out to carry away the drainage water. A developing wet patch of grass is often the sign of a broken land drain

Field shelter for pony

that should be repaired. Other annual maintenance tasks well worth the trouble include harrowing with a chain harrow, rolling, and where possible, mowing the seed-heads off the grass in July to encourage leaf growth well into the autumn. Keep a constant watch, throughout the year, for dangerous litter such as plastic bags, tin cans, and bottles, and for the poisonous plants listed on p. 187.

Good stock-proof fences or hedges are another essential. When fencing is used it should be made of stout wooden post and rail. For safety, never use wire netting, strand wire, barbed wire, nor palings, whether made of wood or iron.

Companions Ponies are by nature herd animals, and ought not to be kept singly. Given the opportunity, they seek the company of other ponies, donkeys—even goats—rather than remain alone. Conversely, it is unwise to keep too many ponies together. This can be one of the disadvantages of hiring grazing, and the dangers are that ponies will be injured by kicking, that the grass will be spoilt by overgrazing, and that the land will become increasingly infected with parasitic worms.

Field Shelter Healthy ponies are able to survive outside all the year in Britain, being descended from our hardy native breeds which grow very thick winter coats,

Stabling suitable for pony or donkey

but they must have some shelter. In the wild they would be free to benefit from whatever natural shelter is available, but there is little to be found in a single field. Good hedges and trees provide some protection from wind and rain, and also from sun and flies, but a pony should also have access to a field shelter. The shelter has to be big enough to accommodate all the ponies using the field. The walls and roof must be weatherproof, but the shelter is open to the ponies at all times with a $2\frac{1}{2}$ m wide entrance instead of a door. It can be closed, by poles fixed across the entrance, when it is necessary to give some protection to a pony that has no loose-box.

Loose-box and Tack-room In practice it will be found unsatisfactory to try to keep a pony without a loose-box and tack-room within easy reach of the house. There are times—if the pony is sick, the weather severe, or the grass too lush for the pony to graze all day long—when the pony is best stabled in its loose-box, with the top half of the door open. If the loose-box is spacious, with adequate head room, good ventilation, and a good deep bed of straw, shavings, or peat, the pony will be comfortable and able to move around a little. This is particularly necessary at night, because horses and ponies sleep only fitfully, and need space to change their position. Ventilation is by means of louvres or open windows, fitted with grilles. Never exclude fresh air from the box. If extra warmth is needed, it is better to use blankets and leg bandages than to shut the window.

Without somewhere warm and dry to store the pony's food, the stable equipment, and yard tools, it can be very difficult to manage a pony well.   A tack-room is needed.

Mucking Out   The loose-box must be 'mucked out' each morning.   First the droppings are removed in a skep or bucket kept for the purpose, and then the soiled bedding.   Add enough fresh bedding to make a good deep bed for the pony, and never leave a pony in its box without a comfortable floor covering.

Ponies will not eat grass fouled by their own droppings, and this is one reason it is necessary to have enough land to use in rotation.   In a more confined space, the remedy is to pick up the droppings from the paddock every day.

Care of Tack   The proper care of tack is important not simply for good looks, but for the comfort of the pony, and the safety of the rider.   When cleaning the tack, always note its condition: loose or broken stitching needs repairing straight away, for it may cause an accident; worn saddles need restuffing for comfort.   After wiping clean with warm water, leather is kept supple by the use of saddle soap or neat's-foot oil.   Dried mud or hairs on the

Tack room

saddle lining, on the numnah, or on the girths will soon rub the pony sore, and must be brushed off before use.

Exercise Clearly all ponies need the freedom to exercise themselves, but the amount of work they can do depends to a large extent on their food intake. **A pony at grass** is not in condition for hard work involving galloping or long days. If it is needed for hard work, it must be specially conditioned over a period of several weeks. Begin to feed corn or pony cubes, and start a carefully graduated programme of exercise. At first the pony should be ridden slowly for very short periods. If the exercise is increased only gradually—and never given within one hour of feeding—the pony's condition will improve steadily, and there will be no damage to its wind or to muscle and tendons that are so easily strained if too much exercise is given too soon.

A stabled pony needs daily exercise, and can be turned out into a paddock for a few hours. In winter, if the coat has been clipped, the pony will need to be turned out in a New Zealand rug, or given a period of more warming exercise—either ridden, or loose-schooled on a lungeing rein. Avoid working a pony within an hour of feeding, and begin and end every ride with a walk. This ensures that the pony is allowed to warm up slowly, and not strain muscles, and that the pony is not overheated when returned to its stable.

Grooming When a pony comes in wet—either because it is overheated, or has been out in the rain—an effort should be made to dry it off to prevent chilling. One recognized method, often recommended because it is both simple and effective, is to put straw on the pony's back for about half an hour, covered with a rug. In winter, ponies expected to work hard have their thick winter coats clipped in order to facilitate drying when they come in wet after strenuous work. Obviously, once they are without the winter coat, they have to be stabled, kept warm during the day by a rug. For extra protection at night, a blanket is put under the rug.

Ponies are groomed not only to improve their appearance, but also their skin, muscle tone, general well-being, and comfort. A stabled pony is groomed hard

Grooming tools

every day. Mud, dust, and excess grease are removed from the summer coat or the clipped coat with the body brush. The mane and tail are also groomed with the body brush. Every few strokes the brush will need to be cleaned on the curry comb—which is never used on the pony. Mud is washed off the legs in summer, but in this climate there is a risk of cracked heels developing if the pony's legs are washed in cold or wet weather. It is perhaps safer in bad weather to allow the mud to dry on, and then to remove it with a brush. After grooming the coat, use a damp sponge to clean the eyes, muzzle, and dock.

A pony at grass is protected by the grease in its coat, and needs no more than light grooming to remove mud. The stiffer dandy brush is used on the thick winter coat. Pay particular attention to the saddle and girth marks, where dried mud would chafe the skin.

The pony's feet need special care. They must be picked out before and after every ride—or any other form of work—and each day the pony is stabled. The farrier will need to visit, or be visited, every 5–6 weeks to 'dress' the pony's feet (i.e. to trim back the growth of horn) or to renew the shoes. If doing no work at all, a pony at grass need not be shod. Even so, the feet soon grow out of shape, and will need the attention of the farrier every 2–3 months.

Feeding The pony has the teeth, and also the long gut of the true herbivore, and is the most perfect example of a creature adapted to live by grinding grass. The molar teeth wear down in such a way that they remain effective right into old age. The gut, measuring about 30 m, is long enough to digest grass and hay without rumination. The stomach, however, is relatively small, which means that the pony should not go long without food.

At grass, the pony grazes for much of the day and the night, as in the wild, but will need hay between October and April when the grass, although it looks green, lacks sufficient nutriment. Ponies also need hay at other times of year—for instance, during a period of drought—when the grass is in poor condition. When several ponies share a field, the hay ration should be put out in several nets—one for each pony and one extra to stop bullying.

A stabled pony is fed 4 times a day, and given a good net of hay at night. The feed can be made up of oats, barley, maize, or pony cubes, with some sliced apple, carrot, mangold, and swede for variety.

Good quality hay—either seed or meadow hay—is sweet-smelling, free from dust and mould, and either green or yellow in colour. It can easily deteriorate and become dark, soft, and fusty, and must never be fed to ponies in such a condition.

Grain can be of almost any kind except wheat, but oats are most often fed. They are very stimulating, and should not be fed to a child's mount without first discovering their effect on that particular pony. Since ponies need plenty of roughage in their diet, all grains are usually fed with chaff (chopped hay or oat straw) and/or with bran. An additional advantage of feeding these bulk foods, which need extra mastication, is that they prevent a pony from bolting its feed.

Pony cubes, also known as pony nuts, are the modern convenience food for ponies. They contain grain, mixed with a bulk feed such as grass meal, and ponies seem to find them very palatable. Whether stabled or at grass, every pony must have access to a salt lick. The salt lick supplements the natural supplies of sodium chloride (common salt) in the pasture, and is necessary for blood formation, digestion, and as an appetizer.

Water A pony has a big capacity for water, and will drink perhaps 4 bucketsful (36 litres) a day. If there is a natural supply of clean water available in the field—for instance, a clear stream—the pony will drink from that. Alternatively, the drinking water will have to be piped out to the field, or carried in buckets.

Water a pony before, rather than after feeding, and provide a supply for the night. Ponies drink after sunset, especially in hot weather. In freezing weather, break the ice on the water early each morning, and avoid using chemicals to prevent its forming.

Handling Ponies are sensitive, highly-strung animals needing patient and gentle handling. Always approach a pony from the side, and speak to it quietly as you do so. Use the aids—the signals from rider to pony—carefully, particularly the reins, for the mouth is sensitive and very easily hurt. It is important to have the very best riding instruction available in your area, for a badly taught rider will ruin the pony's schooling. A good rider will gradually improve it.

Ailments LAMINITIS is an inflammation of the sensitive tissues of the feet, which is particularly painful because it is confined within the horny wall of the hoof. Whether or not laminitis is suspected, any lameness in the pony needs immediate veterinary attention, but even if treated promptly, laminitis can cause permanent damage to the feet. Undue weight on the legs—in fat or pregnant ponies—seems to be one cause. Another is an allergic reaction to overeating lush grass in the springtime.

It is very easy for a pony to sustain CUTS and GRAZES without the owner's knowledge. To guard against the TETANUS organism which gains entrance through such wounds, the pony should have an injection that will give it long term immunization. If you are unsure about your pony's immunity to tetanus, or if a cut may need stitching, contact your veterinary surgeon. Minor cuts, scratches, and grazes should be washed with an antiseptic solution.

Coughs, sneezes, nasal discharge, and laboured breathing should never be ignored. Seek veterinary advice as early as possible, for all may be symptoms of serious respiratory tract ailments such as BRONCHITIS, PNEU-

MONIA, EQUINE INFLUENZA, and STRANGLES. This includes the incurable BROKEN WIND or HEAVES, in which the pony experiences increasing difficulty in breathing, due to deteriorating lung tissue.

Digestive ailments are grouped together under the general name of COLIC—a term that means abdominal pain. Common causes include sudden change in diet, severe indigestion, faulty watering, inflammation or strangulation (i.e. twisting) of the intestine, peritonitis, and worms. The pony shows signs of great distress, although the pain varies with the nature of the colic. Sometimes it is sudden, severe, and intermittent; sometimes it builds up slowly over several hours. The pony will probably sweat, look round at its flanks, and try to get some relief by lying down and rolling, or by kicking at its belly. Send for the veterinary surgeon urgently, and meanwhile walk the pony slowly. Keep it warm under a rug, but do not attempt to give food or water.

When a pony shows signs of debility, the veterinary surgeon may recommend a suitable medicine for worming. The WORMS, in fact, recur, for all ponies have them, but a period of relief gives an ailing pony a chance to recover his condition. Normally worms have little adverse effect on a healthy pony.

It often happens that a veterinary surgeon will recommend that an old pony, in deteriorating health, should be put down rather than suffer longer. In this case, the most humane way is to have the pony shot at home.

The Ass or Donkey   *Equus asinus*

Origins   The donkey, *Equus asinus*, Linné, is one of the most charming and desirable of the pet animals, with a status it has seldom enjoyed since ancient times—when it was also highly prized. As a working animal the donkey was too often scorned as inferior in performance to the horse and the mule, and its stolid temperament interpreted as obstinate and uncooperative. Horses and donkeys were commonly overworked, but the donkey must also have been held in general contempt for both 'ass' and 'donkey' have become synonyms for stupid.

The donkey is believed to share a common ancestor with the other equines—horse, zebra, and onager—but from Pleistocene times it evolved quite independently, influenced by very different geographic factors. It is indigenous to North and East Africa and is, therefore, generally accustomed to higher temperatures, greater aridity, more mountainous country, and sparser vegetation than the horse which evolved under quite different conditions on the Eurasian steppes.

It is thought the donkey was domesticated between 4000 and 3000 BC by the Egyptians or Libyans. From Africa it was introduced into Asia, and later into Europe, but was not established in Northern Europe before medieval times. Its introduction to Britain probably dates from the 9th or 10th century.

Varieties The donkey has been as useful an animal to man as any, yet very little attempt has been made to improve it in Britain by selective breeding. There are no recognized breeds as with horses, but for show purposes donkeys are differentiated, according to size, as miniature (under 9 hands); small standard (9–10 hands); large standard (10–12 hands); and Spanish (over 12 hands). Abroad, certain characteristics—particularly those of size and body marking—have become dominant in different parts of the world, giving rise to geographic races such as the dwarf donkeys of Sri Lanka and Sardinia, which measure barely 8 hands, and the giant ass of Poitou (France) which stands over 15 hands at the withers.

Domesticated donkeys are descended from three races of wild ass: the Nubian, the North African, and the Somali. The Nubian evolved in the mountainous regions of NE Africa where there is a faint chance it still survives. It is characterized by the cross marking, formed by a stripe along the back, and stripes down each shoulder. Most donkeys are descended from the Nubian, and retain the cross marking. The North African race evolved in the Atlas mountains of NW Africa. It became extinct in Roman times, but some modern domesticated donkeys have descended from it, and bear

Jackass (right) with Jenny

its characteristic markings of shoulder stripes and transverse leg stripes. The Somali, which still survives under careful protection, has no body markings, but distinct transverse leg stripes. Some modern African races bear similar markings, and may well be derived from the Somali wild race.

Colours Grey is the predominant colour, but there are also brown, black, and white donkeys. Whole-coloured ones usually lack the great variety of face and leg markings associated with horses, but the muzzle, legs, and belly shade to a paler tone. Broken-coloured donkeys include piebalds, skewbalds, roans, and spotted donkeys.

Description The donkey was domesticated primarily as a beast of burden, and as a draught animal. In these respects—if not as a riding animal—it can vie with the horse, being strong for its size, sure-footed, even-tempered, and easy to feed. As a riding animal it has one advantage over the pony; it is a reliable mount for a very young or nervous rider. Physically, the donkey re-

124

sembles the horse in many ways, but has only 5 loin vertebrae, is usually smaller, has longer ears, a thin tail with a tuft of long hair only at the end, a short, erect mane with no forelock, slender legs, narrow hoofs with no hair on the fetlocks, and no chestnuts (callouses) on the hind legs.

Male and Female   The terms stallion, gelding, mare, colt, and filly are used exactly as for horses, but it is also correct to refer to a donkey stallion as a jack, and a donkey mare as a jenny.   Both sexes have 2 teats.

Breeding   Much of the breeding of donkeys is carried out on donkey stud farms.   Donkeys are capable of breeding throughout most of the year, with the possible exception of December and January.   The females are in season for about 7 days every 3 weeks, and 7 days after giving birth.   The gestation period is longer than that of the pony, averaging $12\frac{1}{4}$ months (340–395 days).   The foals are well-developed at birth, like pony foals.   They are weaned at 5–6 months, and attain puberty at about 1 year.   The colts must then be gelded or separated from females to prevent unplanned pregnancies.   Females should not be used for breeding until 2 years old.   Their life span in captivity is often 25 years, with ages approaching 50 years on record.

Hybrids   Donkeys and horses do not breed together naturally, but they are so closely related that they have been crossed by man for more than 2000 years to produce hybrids, which are themselves always sterile.   The off-spring of the jackass and mare is the mule; that of the stallion and jenny is the hinny.   The hinny, being smaller, and considered inferior to both horse and donkey, is seldom bred.

Field or Paddock   The pet donkey is normally kept at grass, with a field shelter.   Donkeys are more economical feeders than ponies, and it is usually sufficient to provide them with $\frac{1}{2}$ hectare of grazing per head.   The paddock needs safe, gap-free fencing about 110 cm high, good drainage, a water-supply, and a field shelter.   It is best divided into 2 plots which can be used in rotation, and managed as described in the pony section (pp. 114–5).

Shelter Donkeys have won a reputation for their hardiness and endurance, but their North African origins make them more vulnerable to the cold and wet of northern winters than our native mountain and moorland ponies. They are most at risk in damp and wet conditions; this applies particularly to donkey foals.

A donkey's field shelter (see also pony section pp. 115–6) must be both waterproof and draughtproof. It should stand on a concrete apron, extending in front to give dry access. The concrete is best covered with oat or wheat straw, and mucked out each day it is used. The soiled straw can be piled up outside, together with droppings picked up from the paddock, to make a manure heap. It is recommended that the opening should face south or south-west, and be fitted with a half door to give extra protection in severe weather, and if the donkey is sick. The shelter need not be as high as a pony's, but it should be remembered that the owner—and perhaps the veterinary surgeon—needs to be able to work inside in comfort.

Companions Donkeys are social animals. In nature, they live in small family herds of a stallion with several mares and foals. In captivity, two or three donkeys will live together more happily than one alone, or they will give companionship to ponies. Check, however, that the smaller donkeys are not kept away from food by the ponies, and that in bad weather they are not crowded out of the field shelter.

Exercise Donkeys, and particularly foals, need freedom to canter round their own paddock. For this reason they should seldom be tethered. It is sometimes useful to be able to tether a donkey to crop a patch of nettles or brambles, but use only an elderly one who will not fret at the restraint. If a donkey is to be tethered, use a leather collar—the mane will not get caught as it would with a pony—and tether for a short while.

If a donkey is to be exercised by riding, or any other kind of work, it will need to be fed corn. Check the fit of the saddle and bridle, and that the rider weighs no more than 50 kg. Donkeys should never be ridden before they are 2 years old.

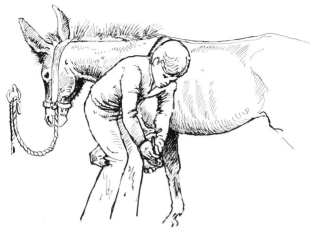

Using a hoof-pick on a donkey's foot

Grooming A donkey's coat needs less attention than a pony's, but the essential grooming is similar (see p. 118). In winter the donkey needs only light brushing to remove mud, but not the protective grease in the coat. In spring, brushing should be vigorous to help the moult of the thick winter coat.

Donkeys kept at grass are not shod unless expected to do road work, but they must visit, or be visited by the farrier every 6 weeks. The hoof is less brittle than that of the pony, and does not break away. Unless trimmed regularly, it can become seriously overgrown causing considerable pain, lameness, and perhaps permanent damage. The donkey's feet must be picked out with a hoof pick, like a pony's.

Food All the equines are strict herbivores, but the donkey, having evolved in North Africa, is suited to a poorer diet than the pony. It is content to graze rougher pasture, thistles, and nettles, and to browse on low-growing scrub such as brambles. Indeed, too much rich food will merely cause laminitis. In summer the donkey can live by grazing unless grass is in short supply. In winter it will certainly need a net of sweet meadow hay each night and morning—hung in the field shelter to keep dry if need be—and probably other supplements if it is to keep its condition. Pony cubes are good, but so too is a mash of bran and oats, mixed with hot water and a little salt.

Clean kitchen scraps such as vegetable peelings add variety to the diet, as do fresh fruit and vegetables— carrot, swede, turnip, apple, cabbage, cauliflower—all of which must be cut into chunks or slices. Donkeys also enjoy a crust of bread, which can be hand-fed to help keep them tractable.

Working donkeys will need to be fed corn—preferably oats—throughout the year. Corn is always given with chaff (see p. 120).

A salt lick (see p. 120) should be fixed in the field shelter, or to a tree, so that each donkey has access to it.

Water The donkey drinks less than the pony, otherwise watering is exactly similar (see p. 121), except that the donkey is fastidious about drinking, and will refuse water that is not clean and fresh.

Donkey with salt lick

Handling   It is said that donkeys are the friendliest of the equines.   They seem to enjoy human companionship, and the owner should be sure to spend some time with them each day.   Those used to human company, and to quiet handling, are generally cooperative.   Others which display proverbial stubbornness have very often not been trained, or have had their early training ruined by rough handling.   The donkey, as well as the pony, needs 'breaking in', but the donkey is much easier to school.   The foals learn very quickly, and should be taught to lead, i.e. to walk on a leading rein when only a few days old, and to continue to do so throughout their life.   Lead by walking level with the donkey's left shoulder, with the leading rein in the right hand.

Ailments   Donkeys are generally hardy, but are subject to all the equine ailments and disorders that afflict ponies (pp. 121–2).   They suffer less from leg and joint complaints, but seem more prone to EQUINE INFLUENZA and to strangles.   STRANGLES is a serious, notifiable disease with symptoms of fever, catarrh, inflammation of the nasal passages, and abscesses in the glands of the neck.

Donkeys are also prone to infestation by LUNGWORMS. These can be passed on to horses, often with serious consequences.   The first symptom is a dry cough, followed by a loss of condition, particularly a wasting of the flesh, and a dry, staring coat.   A veterinary surgeon will prescribe an effective worming powder, which is best fed to the donkey in a bran mash.   Donkeys should also be treated twice a year, or more if necessary, for INTESTINAL PARASITES.   Again, a veterinary surgeon will prescribe a suitable powder.

   If a donkey's health fails in old age, it may be necessary, on veterinary advice, to have it shot with a humane killer at home.

NOTE: In old age ponies and donkeys need devoted care.   Never sell an old pony or donkey.

BIRDS

The Budgerigar   *Melopsittacus undulatus*

Budgerigars were first seen in this country in 1840. Throughout the 19th century hundreds of thousands of wild specimens were netted in Australia and imported to Europe for breeding. They became such popular pets that today they are the most commonly kept of all caged birds; their colour, song, mimicry, sociable nature and easy care make them as popular with the elderly and the housebound as with young pet owners.

Description   Budgerigars are small, colourful parrots with the beak and toes typical of all the psittacine (i.e. parrot-like) birds. In nature they are climbers; the strong beak is used not only for dehusking seed, but as a climbing aid, and the zygodactyl toes—two pointing forward and two backward (below)—are the main adaptation for climbing. The ideal length for a domesticated specimen is said to be 216 mm; the natural stance 30° from the upright; the most distinguishing feature the necklace of six throat spots that shows on many of the colour varieties. Breeders pay much attention to body outline, shape of head, position of wings, and so on, but such considerations need not trouble the owners of pet budgerigars.

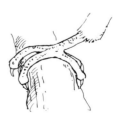

Zygodactyl
toes of
budgerigar

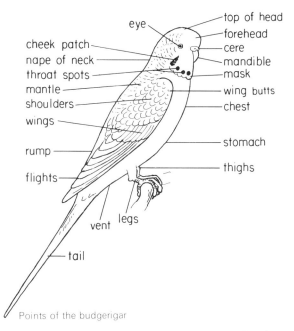

eye
top of head
forehead
cere
cheek patch
mandible
nape of neck
throat spots
mask
mantle
wing butts
shoulders
chest
wings
rump
stomach
thighs
flights
vent   legs
tail

Points of the budgerigar

Natural Colour The wild budgerigar is smaller than the domesticated, and its natural colour is light green on a yellow ground, with black throat and wing markings. The characteristic undulating wave pattern of the wings and the striped head markings are alluded to in its common names of undulated parakeet and zebra parakeet. Ornithologists report seeing colour mutations—particularly light yellows and dark greens—among the wild flocks, but as the natural light green is genetically dominant to all other colours, the mutants have little chance of becoming established in the wild.

Varieties When a mutation occurs in captivity, the new factor can be 'fixed' by breeding the mutant back to its own offspring. The first mutant colour to appear in captivity was the yellow, in the 1870s; the blue was much later in the early 1900s; the white later still in 1920. There are 4 colour series: green, yellow, blue, and white. The wide range of colour is partly due to the presence or absence of the so-called 'dark character', that is particularly noticeable in the greens and blues. This allows for three tones of each of the basic colours:

light green	light yellow	sky blue	white sky blue
dark green	dark yellow	cobalt	white cobalt
olive green	olive yellow	mauve	white mauve

The light green, light yellow, sky blue and white sky blue have no dark character; the dark green, dark yellow, cobalt and white cobalt have one; the olive green, olive yellow, mauve and white mauve have 2. These are the normal varieties, but breeders have produced many more, using 3 other colour characters that have appeared: the grey, slate and violet. Yet further variety is gained by mutations which eliminate or change the pattern of the markings—opalines, clearwings, cinnamons, greywings, lacewings, pieds, and so on. There are also mutations that give different eye colours, and even crested heads. The resulting permutations give a wide range of varieties, but no red or pink budgerigar. A red factor has never been reported and the attempt to breed a red from existing stock is virtually a genetic impossibility.

Betcherrygah The name 'budgerigar' is said to be a corruption of the aboriginal name 'betcherrygah'. Shaw classified these birds as *Melopsittacus undulatus* (Gr. *melos*, a song; *psittakos*, a parrot; L. *undulatus*, wave-like), and the first to be seen in England was a stuffed specimen shown to the Linnean Society in 1835. It is John Gould, however, the 19th-century ornithologist famous for his study of Australian birds, who is credited with being unwittingly responsible for the cult of the budgerigar, for it was he who brought the first live pair to London in 1840.

Natural Habitat The budgerigar is a migratory flock-bird of the Australian interior. Several species are common, *Melopsittacus undulatus* being reported in the grasslands of Queensland, New South Wales, and parts of South Australia. The inland plains are semi-arid grass-

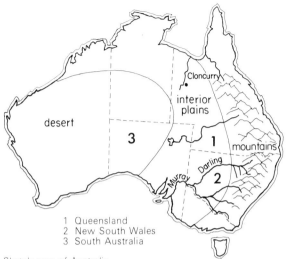

Sketch map of Australia

1 Queensland
2 New South Wales
3 South Australia

lands with an annual rainfall below 50 cm, supporting only coarse grasses and stunted eucalypts (gum trees). The birds are dependent on both. Seeding grasses provide food; eucalypts provide shade, green food, roosting places and nesting sites.

Adaptations to Natural Habitat In order to avoid the twin ravages of drought and excessive heat the budgerigars need to live a nomadic life, constantly moving on in search of fresh pastures and water-holes. In much of their territory artesian water is available at the surface, and the birds flock from one water-hole to the next, but although budgerigars are economical in their use of water, and will fly long distances to find a fresh supply, there is still a high mortality rate in times of drought. The birds are also constantly seeking fresh feeding grounds. Rainfall is low, but when it does fall temperatures are high enough to ensure very rapid seed germination, and short-lived but lush pastures grow up in a matter of days. The budgerigars' roving habit, that allows them to locate these temporary pastures as they appear, is essential to survival in a semi-arid environment.

Budgerigars are reported to rest in the eucalypts during the heat of the day, but tropical heat is more effectively avoided by annual north-south migration. The birds do not return to the same place each year, but the general pattern of migration is that in the spring the birds fly south to avoid the tropical summer (Cloncurry, Jan. 30°C), returning in the autumn to take advantage of more comfortable winter temperatures. An exception is reported to occur in an unusually dry summer when the birds fly north, in spite of the high temperatures, to feed on temporary pastures that grow up following violent tropical thunderstorms.

The birds breed in the summer, from October to December, while in the south, one of the favourite areas is the mallee scrub of the Murray river basin. The mallee is a species of eucalypt with a distinctive habit that favours the breeding budgerigar population. Many trunks—hollow, gnarled and knotted—grow from each rootstock to form a dense scrub of dwarf eucalypts, about 2 m high. The budgerigars lay their clutch in any convenient hole, using no nesting material whatever. This scrub is an ideal habitat for the breeding budgerigars; not only does it provide countless nesting sites for the huge flocks, but food as well. During the winter there is sufficient rain for lush kangaroo grass to grow up, carrying enough seed to satisfy the huge migrant population.

Life Cycle in Captivity A clutch normally consists of 3 to 10 white eggs laid on alternate days, the chicks hatching in succession after 18 days' incubation. The young are hatched weighing about 2 g, the birth weight normally increasing to 40 g in 21 days. In the nest the chicks are entirely dependent on the hen who feeds them on regurgitated seed and so-called 'crop milk' which is particularly rich in protein.

At hatching the chicks are blind and naked; eyes open and feather growth begins at 6 days; at 14 days they are able to move about. Wing and tail feathers are formed by 21 days, the plumage being complete by 28 days when the chicks are ready to leave the nest box. At this stage the heads of the young budgerigars are marked with dark horizontal bars of colour which disappear during the first moult at about 12 weeks. Most young budgerigars also have black eyes. The light-coloured ring around the eye occurs only in adults. Budgerigars are fully fledged at 5 to 6 weeks, and sexually mature at 4 months, although much too young for breeding, which should be delayed until they are nearly a year old. Life span in captivity is normally 5 to 8 years, but many budgerigars live for 10 years, and ages as high as 21 years are recorded.

Male and Female The male is known as the cock; the female as the hen. Adult weights are between 35 and 60 g, the average being in the range 40 to 45 g. The male is often a little bigger than the female, and body outline is slightly dissimilar in that the male shows a higher domed head. Budgerigars can usually be sexed by the colour of the cere—the skin above the beak. As a rule the cere of the male is a smooth blue or violet; that of the female is brown and rough in texture. Exceptions occur among juveniles in whom the cere colour is indistinct, and in adult males who may lose their cere colour with ill-health or with age.

Breeding As budgerigars are flock-birds, they should never be kept singly; it is much more humane to keep two. A true pair will breed readily, the cock taking on a great deal of the parental responsibility. He feeds the hen during the 18 days of incubation—when she leaves the nest for no longer than a few minutes each day—so that the satisfactory development of the young depends on the cock's ability to feed the mother. At the end of the nursing period the cock is occupied for some days helping to feed the fledgelings who, to begin with, are slow at dehusking seed for themselves. The cocks, therefore, as well as the hens quickly become exhausted by repeated breeding, and should be limited to rearing no more than 8 to 10 chicks in a season.

This number of chicks will probably result from 2 or 3 clutches. Clutch size varies considerably, but the first clutch of the season is likely to be larger than later ones. As many as 8 eggs are not uncommon in a first clutch.

Pair of
budgerigars
with young

Subsequent clutches vary considerably. It may be
necessary to separate the pair to prevent further breeding,
but very often it is sufficient just to remove the nest box.

Nest Boxes Budgerigars do not build nests in the wild,
and will not do so in captivity. Birds in breeding
condition will lay eggs and attempt to rear a clutch in any
available hole, such as one between the outer and inner
walls of an aviary, and a hen in breeding condition will
simply drop eggs from her perch. For these reasons
boxes must be provided for nesting.

In the past, hollowed logs of birch or poplar were used,
but nowadays boxes such as that shown opposite are
more common. Some are constructed of wood; others
are made of hardboard with a wooden tray. A good
design allows for easy inspection of the clutch without
disturbance of the birds, and also for easy cleaning. The
minimum size is 17 cm × 17 cm × 23 cm high, and the
entrance hole should be about 4 cm in diameter. It needs
to be placed off-centre—not immediately in front of the

nesting hollow—so that the hen does not damage the eggs when entering. A perch just outside the entrance hole allows the cock to feed the hen during the incubation period. The wooden bottom should be in the form of a removable tray, with a saucer-shaped depression about 14 cm in diameter hollowed out as a nest. A glass inspection door is recommended as it allows the owner to examine the nest without the chicks becoming chilled, or accidentally pushed out. Ventilation holes are incorporated to avoid oxygen starvation of the young when the hen sits in the round entrance hole and blocks off that air supply.

The eggs are laid directly on to the wood without the use of nesting material. Most owners keep the hollow covered with a sprinkling of fine sawdust to facilitate cleaning while the chicks are developing. During this month the nest box may be scraped clean of droppings, and the beaks and feet of the chicks cleaned with damp cotton wool if necessary. It is safe to handle the chicks, and to transfer them to a cardboard box while the nest box is being cleaned.

Breeding Condition Budgerigars can reproduce at any time of year if they are in breeding condition, but the breeding season usually extends from February to September, the birds resting during the winter months. In breeding condition the cere colour is very

Box-type nest box

pronounced. The plumage is unlikely to be in show condition, but a bird should not be paired if it is moulting. In addition to the strong cere colour, the cock will be very perky, and make advances to the hen, probably feeding her with regurgitated food. She too may regurgitate food—which should not be confused with vomiting—will respond to his calling, and she may drop infertile eggs from her perch. She will also gnaw at any available woodwork, showing great interest in possible nesting sites.

Pairing The recommended breeding ages are within the limits of 10 months to 6 years for cocks, and 11 months to 4 years for hens. The normal practice is to pair up birds of equal merit: the best hen to the best cock; the second best hen to the second best cock and so on, but not if a particular pair shows the same fault, e.g. crossed wing-tips. If the chosen pair will not accept each other, the only course is to offer a different partner.

Cage and Colony Rearing It is normal to house a single pair in a breeding cage fitted with a nest box, but beginners should not provide the nest box too early in the year. Although a fancier may breed early in order to have birds ready for showing later in the same season, the pet-owner can wait until mid-March. Should an owner wish to breed from a bird born late in the previous season—in August or September—he would be well advised to wait until June, when it is 9 or 10 months old, and then to allow one clutch only that season.

When several pet birds are kept in an aviary, colony rearing is possible. Clearly such an indiscriminate method would never be used by the serious fancier, concerned with breeding for colour and for quality, and with keeping accurate records, but it is an acceptable method of breeding pet birds. Indeed it is even possible to exert some control over the results if, a week or fortnight before the nest boxes are introduced to the aviary, a selected pair is removed and caged together for that time. The faithfulness of the cock to a particular hen cannot be guaranteed once he is returned to the aviary, but it is likely that he will pair up with her, especially if he began to feed her in the cage. It is not good management to have a

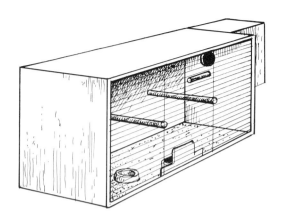

surplus cock in a breeding colony.   When the number of cocks and hens is unequal an odd hen is not likely to be difficult—one of the cocks will probably rear two families simultaneously, but an extra cock, desperately looking for a mate, can be very troublesome.

It is best to allow 2 boxes per pair, since budgerigars seem to appreciate having a choice, and the boxes should be placed at the same height, or there may be squabbling among the hens for the possession of the highest, which is invariably the birds' first choice.   If 2 hens fight for the possession of a particular nest box, it is best removed. Very often both will quickly select an alternative.   The height at which the boxes are fixed need only be limited by the reach of the owner; it should be possible to see inside without first having to remove a box from the wall.

Housing in Cages   The sight of a single budgerigar isolated in a small cage—although common—can be so distressing that one might be forgiven for condemning outright the caging of budgerigars, but providing they are given daily exercise and have plenty of human contact it is possible to house them in cages.   Indeed it is the only method for housebound people to use, many of whom are excellent owners and devote hours to training and talking to their budgerigars in return for their companionship.

The first necessity is space. The recommended size of a cage for 2 budgerigars is 100 cm long, 60 cm deep, and 75 cm high. By choice the cage should be made of wood with a mesh front—either weldmesh or aviary cage fronting. If aviary cage fronting is used, a design with horizontal bars is preferable as it allows the birds to use it as a climbing frame. A wooden cage is much less draughty than an open cage and—with the addition of a nest box—can double as a breeding cage. The best position is out of reach of cats and dogs, in a light and airy room where people do not smoke. Beware of draughts, and of direct sunlight, which can cause sunstroke. Budgerigars need to sleep during the hours of natural darkness, so at night the cage will need to be covered if the room is lit—if only fleetingly by the headlamps of passing traffic—and the cover removed early next morning.

Essential accessories for the cage are containers for seed, grit and water; mineral block and cuttlefish 'bone'; sand or sanded paper for the cage bottom; and a selection of perches. There should be a perch in front of each container, and high perches so positioned that droppings cannot contaminate food or water. Most bought cages have 4 dowelling perches of exactly similar diameter, but leg and foot muscles tend to become cramped if always in the same position, and so perches of varying thickness, or branches are recommended. A manufactured cage can usually be improved both practically and visually just by removing the top perches and replacing them with a branch cut from a fruit tree. Another advantage of natural perching is that claws are worn down naturally when the budgerigar can get a firm grip around a branch.

Although it is a mistake to clutter a cage with accessories, some toys are necessary, particularly for a lone bird. All budgerigars seem to appreciate a mirror. A small one inside the cage is useful, but is probably not as satisfactory as a larger wall-hung mirror close behind the cage. Other favourite toys include ladders, bells, swings, and hanging ping-pong balls. It is natural for a budgerigar to peck and gnaw, but for safety a plastic toy should be replaced once the bird has succeeded in breaking its surface.

Exercise   Caged budgerigars need a period of free flight every day.   Once the room has been made safe—paying particular attention to the dangers of electric fires, cats, dogs, fire-places, open doors and windows—the cage door should be clipped open and the budgerigars given at least a 20 to 30 minute playtime.   A budgerigar which is used to its cage, and reasonably tame, will return to its home unbidden once it has had a period of exercise, particularly when it is allowed out of the cage regularly.

Cleaning   Budgerigar accommodation requires frequent cleaning.   Droppings very quickly accumulate, and together with seed husks make a litter that needs removing from a cage daily.   The actual cage—together with all its accessories—needs cleaning thoroughly twice a week.

Housing in Aviaries   Budgerigars may be housed in an aviary throughout the year in this country.   They are hardy enough for a breeding colony of feral budgerigars to have survived for years in the Scilly Isles, and other colonies are reported farther north from time to time, although they invariably die out in a severe winter.   In order that aviary birds may safely overwinter, their accommodation must be sturdily built.   The obvious advantages of an aviary are freedom for these flock-birds to associate together, and freedom for them to fly, although even in an aviary their flight pattern is very limited.   Compatible aviary companions include cockatiels, weavers, and zebra finches.

Aviaries vary considerably in design, particularly as they are often adapted from existing sheds or outbuildings, or built on the lean-to principle, but typically they consist of a warm, dry, enclosed sleeping area adjoining an outdoor flight area.   At night the birds are shut off from the flight by closing a small connecting hatch which they use as an entrance.   The sleeping quarters are ventilated, and are provided with perches for roosting; the flight area is furnished with perches, a feeding table and, when need be, nest boxes.   Food and water is given in the flight area which is partly roofed to provide a dry, sheltered feeding place in any weather.   The aviary needs to be so positioned that it receives the sun at some time of day,

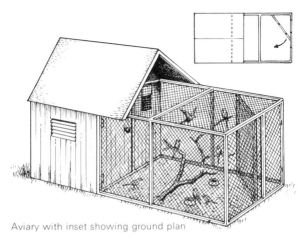

Aviary with inset showing ground plan

preferably facing south-east or south-west. North-facing aviaries are not recommended in this climate.

Safety precautions have to be taken to keep rats out and budgerigars in. To provide vermin-proof accommodation for their birds, some owners use a cement floor and well-fitting small mesh wire; others retain an earth floor, but sink a fine mesh wire barrier well down into the ground around the whole perimeter of the aviary. An entrance porch with 2 doors will prevent the possibility of escape if the rule of closing one door before opening the next is adhered to strictly.

Feeding The staple food of budgerigars in captivity, as in the wild, is seed. The traditional mixture is of canary seed and millet, but such a combination has deficiencies and might prove inadequate for birds in need of a really good diet, for instance at breeding time. Canary seed and millet both have a high carbohydrate content, and the balance of the mixture is much improved by the addition of an oil seed, e.g. red rape, linseed, or niger which has a high fat and protein content, and introduces essential vitamins and minerals. The basic mixture can be bought—clean and packeted—under a brand name such as Trill, which has the necessary additional nutrients in the artificial 'sunshine grains'. Owners of just 1 or 2 pet budgerigars are advised to use the best proprietary brands for convenience, although fanciers may prefer to make up their own supplies with the 2 basic seeds mixed by weight in the proportions $\frac{1}{3}$ canary seed, $\frac{1}{3}$ white millet, $\frac{1}{3}$ yellow millet *or* $\frac{2}{3}$ canary seed, $\frac{1}{3}$ white millet. The second mixture is more expensive, but better than the first, for canary seed is considered more nutritious than millet; white millet is reckoned better than yellow, and yellow better than brown. Senegal millet, sold as millet spray, is a very good variety of that seed, but with its high carbohydrate content should be offered only in moderation if the birds are in danger of becoming overweight. Other seeds, e.g. sunflower and wheat germ, can be fed in addition to the basic mixture, and seeding grasses provide welcome variety for the budgerigars. They are tied in a bunch and hung up for the birds; other seeds are offered in dishes or hoppers, but it is important to realize that budgerigars de-husk seed before swallowing the kernel, and a deposit of husks builds up on the surface of the seed container. This deposit can completely bury the whole seed, and budgerigars have been known to starve because of it. Owners need to make a habit of blowing away the husks every time they replenish the seed supply.

Fresh green food, fruit, and vegetable is appreciated by budgerigars at all times, which is the reason why the earth and shrubs in aviaries are usually bare. Deciduous branches (but not the poisonous laburnum), grass turves, chickweed, dandelion, groundsel, salad greenstuff, apple and grated carrot are all recommended. In a cage, these are clipped to the bars or placed in dishes; in an aviary they are hung in a wire basket. Many budgerigars have conservative tastes and take time to accept changes in their food. It is worth persevering, however, to give them the advantage of a good, well-balanced diet.

Two very important supplements are grit and cuttlefish 'bone'. Seed-eaters such as budgerigars frequently take grit before food. It grinds seed in the bird's gizzard, and is an essential part of the digestive process. Cuttlefish 'bone' is a valuable source of calcium, and is also used by young budgerigars to trim their beaks. It should always be available, firmly fixed into position in such a way that it will not 'give' when the bird uses it. Even so, some budgerigars will ignore cuttlefish if they find an alternative such as mortar which they can peck from a wall. A mineral block should also be supplied.

Water for Drinking and Grooming Budgerigars usually drink only small amounts of water, but it must be available at all times, changed daily, and provided in a clean container. It is necessary for preening, as well as for drinking; unless a budgerigar can wet its feathers it cannot preen properly. Some birds will use a bird-bath or saucer of water; others prefer to roll in a wet turf; most aviary birds drench themselves in the rain. If these methods fail, it is possible to spray a budgerigar lightly with an atomizer spray. Choose a fine day, and make sure the feathers have time to dry thoroughly before the bird goes to roost.

Talking Budgerigars are most amenable to learning when very young, and any training given should begin as early as possible—at 5 or 6 weeks—as soon as they are independent of their parents. At this age a budgerigar with the right aptitude is ready to begin to learn to talk; by 6 months it is likely to be too old. As a rule cocks are better mimics than hens, and women—because of their higher pitched voices—better trainers than men.

The birds succeed in talking only by reproducing exactly what they hear. Although the appropriateness of some of their chance remarks is devastating, it is not suggested they understand words, so any taught must be said clearly, always with the same inflexion. Short, distinct words should be attempted—beginning with the bird's own name—before a phrase or multi-syllabic word is taught. As it may take weeks to teach the first word, patience is the main attribute of a good trainer, who knows she is wasting her time unless she has the full attention of her pupil.

In order that she may get that attention, a bird in training has to to be caged alone, and come to rely for companionship only on its trainer, who speaks to it perched on her finger, or at least sits very close to the cage while she repeats the selected word, over and over again.

Finger tame budgerigar   Normal handling method

If the chosen budgerigar is a suitable candidate for learning to talk, the trainer will be rewarded one day by suddenly hearing the bird trying out the first word of its vocabulary. Only then is it wise to progress to the next word.

Handling Budgerigars are sturdy enough to tolerate frequent handling, and should be thoroughly used to it before being allowed free flight out of a cage. It is also very convenient to have a bird 'finger tame'. A little while spent with a young bird, gently encouraging it to perch on the finger, is time well spent. Tame birds invariably return to their cage without fuss after exercise; wilder ones may need catching. Often they will perch on a pencil or stick held towards them, and allow themselves to be transported back to the cage in that way. At night they can be caught when perching somewhere accessible just by turning out the light. They will not attempt to fly in the dark, but there will almost certainly be enough residual light in the room—perhaps from a TV set—to allow the owner to see. The accepted way of picking up a budgerigar is with a hand over the back. The standard way of catching an uncooperative or panicky budgerigar in an emergency is to throw a duster or cloth over it.

Ailments The most serious budgerigar disease is probably PSITTACOSIS, or parrot disease. This infectious disease is caused by a virus that can be carried by all psittacine birds, and is particularly dreaded because it can be transmitted to man. Established pet birds are not likely to become infected, but the disease is now more common than 20 years ago, and rapidly spreads through aviaries of breeding birds. For this reason alone it is necessary to buy budgerigars only from known, reliable sources. The symptoms of psittacosis are a greenish diarrhoea, nasal discharge, and a general malaise. Prompt veterinary advice must be sought, rigorous standards of hygiene enforced, and children allowed absolutely no contact with a suspected case. Antibiotic treatment is successful both with birds and with man, in whom the main symptom is a recurring fever.

The budgerigar has no associated parasite, although the RED MITE, *Dermanyssus gallinae*, will sometimes affect it. The mites breed in crevices in cages and nest boxes and, if found, indicate that the standards of hygiene are too low. The mite can be killed off with an insecticide; the bird itself may be dusted with pyrethrum powder, or sprayed with a pyrethrum aerosol.

A grey encrustation of the face, feet, or legs is SCALY FACE—a highly infectious condition caused by a microscopic insect. The affected bird must be isolated and treated with one of the proprietary creams sold for the purpose. Complete cure is possible. All accommodation and accessories have to be scoured with disinfectant.

Fever, loss of appetite, and a pumping action of the tail are the symptoms of COLDS, which can deteriorate into PNEUMONIA. Increase the room temperature to 27°C and keep it constant. Recovery depends to a large extent on encouraging the bird to feed again, so offer any known favourite titbits.

An OVERGROWN BEAK or CLAWS can be clipped back. Great care must be taken not to cut into the blood supply, but in case of accidentally causing bleeding, use an antiseptic. If the beak is damaged, provide soft food— bread and milk will suffice—for a few days in case it is too sore for the budgerigar to de-husk seed.

Some chicks, known to fanciers as 'runners', leave the nest unable to fly because they have dropped at least their tail and flight feathers. This is FRENCH MOULT, and as it may possibly be passed on genetically, it is not advisable to breed from strains in which 'runners' appear. The causes of the condition are not known, but several theories are put forward, including that of oxygen starvation in the nest.

If a budgerigar has to be transported to a veterinary surgeon it is best moved in its own cage, covered with a cloth. A really sick bird, unable to perch, can be taken in a small cardboard or wooden box. If it is necessary to destroy the bird, a veterinary surgeon will do so painlessly.

The Canary   *Serinus canaria*

Many foreign finches are kept as pets in Britain, but the canary maintains its position as the most popular. Its success as a pet and fancy bird is enhanced by its unique reputation in mining communities, where it is held in some esteem. This dates from the years when it was used underground by miners who had to risk exposure to several deadly gases. The canary so quickly succumbs to carbon monoxide that it was commonly used as an indicator of the presence of gas and, to this day, the canary has a place in mine rescue teams.

Description   The wild canary, *Serinus canaria* (Vieill.), is a greenish-yellow finch about 11 cm long, undistinguished in looks, but notable for its remarkable song. Finches are passerine (i.e. sparrow-like) birds. Unlike budgerigars they are perchers, with the familiar arrangement of three toes pointing forward, and one backward.

147

Yorkshire canary

They are like budgerigars in being seed-eaters which need to de-husk seed before feeding on the kernel.

Origins The canary is a flock-bird, indigenous to the forests of the Canary Islands, from which it takes its name, and to the Azores and Madeira.   It was spread throughout Europe, as a captive bird, by the Spanish after their conquest of the Canary Islands in the late 15th century. The birds were in such demand that the Spanish developed a thriving trade, keeping the monopoly by the simple ruse of selling only the male birds.   The Spanish monopoly was not broken until, in 1622, a ship carrying a cargo of canaries foundered in the Mediterranean and many of the birds escaped or were released.   They were blown towards Elba where they established a strong feral colony that allowed the Italians, and later other Europeans, including the British, to begin canary-breeding.

Varieties The Germans developed the Roller canaries, with a distinctive song unlike that of other canaries.   The

Dutch concentrated on plumage, raising a variety with curled feathers; these birds were known as Frills. Other distinctive varieties are the Lizard, an old British breed which gets its name from the scaly pattern produced by the well-defined edges of the feathers, and the crested variety.

The three leading British strains are the Border, the Yorkshire, and the Norwich. The Border is a small, prolific canary, not longer than 14 cm. The Norwich is a heavier bird about 17 cm long, and not quite so easy to breed. The Yorkshire is considered the most elegant of the canaries, being long and slim, a good male specimen measuring perhaps 25 cm. All are hardy, and can be raised successfully by beginners.

Colour The colour range of canaries includes white, buff, yellow, green, cinnamon, orange and red. Breeders place such emphasis on colour quality that exhibitors have to 'colour-feed' show birds with special foods or preparations. Some colours, such as white and cinnamon, are the result of mutation. Others, including the much sought red, are the result of hybridization. Canaries have always been bred with other finches (although usually producing sterile offspring called mules) and breeders have been able to introduce a red factor by successfully crossing the canary with a South American finch, the hooded siskin, *Spinus cucullatus*. As yet, red canaries are uncommon, but it is likely that red varieties of the different strains will be bred in time.

Life Cycle in Captivity A clutch normally consists of 4 (1–6) speckled, blue-green eggs, laid at 24-hourly intervals. Most breeders believe it best if the eggs hatch on the same day. They ensure this by replacing each fertile egg with a china substitute, returning the real eggs to the nest once the third has been laid. The eggs are stored in a box at room temperature to delay their development. The incubation period is 14 days.

The nestlings are reared on soft food that is fed to their parents who regurgitate it to the young. The normal adult diet of seed and greenstuff is withheld for the first 10 days. Instead the birds are given, throughout the daylight hours, 4-hourly feeds of a proprietary rearing food or a mixture of egg yolk, plain biscuit crumb, and baked wholemeal bread, fed moist and crumbly. If hard-boiled egg is used, moisten the mixture with a little fresh milk. Seed is gradually reintroduced as the young begin to leave the nest and feed themselves.

At hatching, the nestlings are blind and naked, except for a soft down, but the eyes open and feathers begin to grow during the first week (7th day). The young begin to leave the nest at 3 weeks, and are fully feathered in 4 weeks. Most are quite independent at 5 weeks, but if they experience difficulty in eating seed, some soft food may be continued for a while, and seed can be soaked for two or three days to soften the husk. Occasionally canaries are very long lived, reaching about 20 years, but their normal life span in captivity is 5–6 years. Their breeding life extends from 1–3 years.

Moulting A canary's first moult takes place when it is 6–8 weeks old, and the juvenile body-feathers are shed. The full moult lasts about 6 weeks and takes place in July and August in this country. During the full, annual moult, all the plumage is shed—including the flight and tail feathers. Birds are 'unflighted' until they have gone through a full moult, after which they are known as 'flighted'. Those born too late in the season to

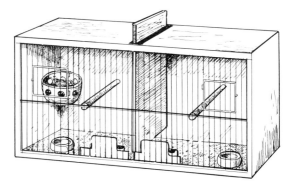

Double breeding cage

experience that year's full moult may therefore remain unflighted for 10 months or so. Canaries need extra care to help them through the moult which places such a strain on the birds that the cocks stop singing. Keep the birds in a warm, even temperature, away from draughts, and feed well. Proprietary moult foods are available.

Male and Female The male is known as the cock; the female as the hen. Adult weights vary according to breed within the limits 12–29 g, the male of a breed normally being longer, and weighing a little more than the female. The sexes are best distinguished by their song. The male is the songster; the female makes only a quiet cheeping sound.

Breeding Canaries are bred only in the spring, either in March or early April, when the temperature reaches 10°C. The usual practice is to pair up two canaries in a double breeding cage (p. 150), the cock in one compartment, the hen in the other, separated initially by a solid partition. The recommended cage size is 100 cm × 30 cm × 45 cm high. After a few days the solid partition is replaced with a wiremesh. When both birds are in breeding condition they will call to each other, and the cock will be seen to feed the hen through the partition, which can then be removed. As the hens are very aggressive at breeding time, these precautions are necessary to protect the cock.

One clutch a year is normally considered enough for a young canary; an older bird may be able to raise two or even three clutches. If more than one clutch is to be attempted, the hen must be provided with another nesting site and nesting materials when her first clutch is 16 days old. Control the numbers raised by destroying surplus eggs.

Nest Pans Canaries are nest-builders, and must be given nesting materials such as hay, moss, grass, feathers, felt and cottonwool. They construct the nest in a nest pan, fixed at the back of the breeding cage, at a height that will allow the hen to feed her young once they hatch. Allow enough headroom for the hen to perch on the rim of the nest pan when regurgitating food to the young.

The hen should not be disturbed during incubation or

Nest pan for canaries   Birdbath for budgerigars and canaries

rearing, or there is a danger she will desert her eggs or her young. Breeding cages have removable base trays which should be cleaned, but the cages are not otherwise disturbed. The cock is left with the hen unless he interferes with the incubation. If he has to be moved, he should be reintroduced by the time the young are a week old, to help with their rearing.

Colony Rearing Colony breeding in aviaries is frowned on by fanciers since it is impossible to control or to record accurately, but it is sometimes used by pet owners. The recommended ratio is one cock to every three hens. Provide more nesting pans than will be used, to allow for individual choice.

Housing in Aviaries Canaries are best housed together in aviaries but, unless breeding, keep only one sex. The single species colony is easiest to manage, but canaries will often live compatibly with other species of similar size and habit if the accommodation is spacious enough. Java sparrows and weavers would normally be suitable aviary companions for canaries. The highly amenable zebra finches and the fire finch are also often put with canaries. Never attempt to keep canaries with budgerigars, which would bully them. When introducing newcomers to an aviary, do so early in the day to give them plenty of time to settle in before roosting. It is wise to supply extra feed pots and perches at the same time, because established birds are likely to be territorial over the existing ones. For information on aviaries refer to the budgerigar section pp. 141–2.

Housing in Cages   If canaries have to be kept indoors, a double breeding cage with the partitions removed makes a satisfactory cage for a pair.   Further information on cages is given in the budgerigar section pp. 138–40.   As perching birds, canaries do not need horizontal cage-bars for climbing, and they usually show less interest in toys. It is not always possible to give canaries free flight in the room; some panic when let out.   In this case they must remain caged always, but use a cage large enough to allow some flight, or preferably re-house in an aviary.

Feeding   The adult diet of seed and greenstuff is supplemented with grit, cuttlefish 'bone', and a mineral block as described in the budgerigar section on pp. 142–4. The only difference is that canaries must have a seed mixture with a higher fat content.   This is supplied by the introduction of niger, hemp, linseed, or more usually red rape.   For preference use a good proprietary mixture such as Canary Trill, or make up this formula, by weight—$\frac{1}{4}$ white or yellow millet, $\frac{1}{2}$ canary seed, $\frac{1}{4}$ red rape.

Water for Drinking and Grooming   (See also budgerigar section p. 144.)   Canaries are very much more dependent on water than are budgerigars, and clean water for drinking and bathing must be supplied every day. Canaries die within 48 hours if unable to reach drinking water.   They like to bathe much more than budgerigars, and for them a bath (p. 152) is a necessary cage accessory.   Always remove the bath early enough for the birds to dry before going to roost.

Handling   Pet owners are warned that canaries cannot tolerate much handling without suffering great stress.   If necessary, handle as the budgerigar (pp. 145–6).   Use a butterfly net to catch an aviary bird or an escapee.

Ailments   The RED MITE (see budgerigar section p. 146) is often found on canaries, and can cause anaemia and death, unless eliminated.   The other most likely canary illnesses are the RESPIRATORY DISORDERS (see budgerigar section p. 146) with symptoms of fever, loss of appetite, laboured breathing, and a pumping action of the tail. Prompt veterinary help is essential.

REPTILES

Mediterranean Tortoises
Testudo graeca and *Testudo hermanni*

The plodding, well-known tortoise is predictably the most popular of the pet reptiles. But the tortoise—unlike many of our exotic pets nowadays—was never bred here, but imported. This trade in wild tortoises, which dates back nearly 100 years, has recently been stopped by an EEC regulation.

Varieties Although no longer for sale in the pet shops, two species of Mediterranean tortoise are still kept as pets in Britain: the Greek tortoise, *Testudo graeca*, Linné 1758, which has no particular association with Greece, and is sometimes called the Mediterranean spur-thighed tortoise; and Hermann's tortoise, *Testudo hermanni*, Gmelin 1788, sometimes called the Mediterranean spur-tailed tortoise. As these common names suggest, the one is identified by a claw-like spur on each hind limb and the other by a similar claw at the top of the tail.

Description The upper shell is known as the carapace; the lower as the plastron. They are joined at each side by a bridge formed by an extension of the plastron. The shell is made of bone, covered with horny plates or shields,

v = vertebral
c = costal
m = marginal
n = nuchal
s = supracaudal

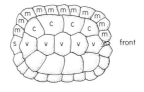

Carapace of
Mediterranean
tortoise

154

Greek tortoise: in twelve years nearly two million were imported for the pet trade

arranged as shown. There are always 5 vertebrals and 8 costals, but the number of marginals is variable. In both the Mediterranean tortoises mentioned there are 23 marginals, although the supracaudal is divided in the male of the spur-tailed (Hermann's) tortoise, apparently giving it 24. The number of shields is constant throughout a tortoise's life, and new growth rings form as its size increases. It is misleading to assume that the rings are annual rings of growth, especially as the ridges tend to become worn down with age, but they give some indication of age to biologists who have studied them.

Although there is considerable variation among members of the same species, in general the spur-thighed (Greek) tortoise has the familiar pale brown shell with dark markings; the spur-tailed (Hermann's) tortoise has a higher domed, smoother shell, often yellow in colour, and sometimes with striking black markings. Against the green lawn of a garden in this country such shells can show up most conspicuously, but their purpose in nature is almost certainly that of protective coloration. Distinctively marked shells have the effect of blurring the outline of the tortoise among the sands, rocks, woods, or scrub of the natural habitat, so providing an effective camouflage against predators.

Biologists also suggest that the high dome of a tortoise's shell is a natural protection against enemies, since it would be very difficult to hold in their jaws. (Significantly, perhaps, most terrapins lack these high domes, and are able to lie flat and unseen at the bottom of a pond.) A tortoise, too, has protective scales on the limbs and tail, but most of all it relies for safety on its remarkable shell into which it can withdraw at the first hint of danger.

Evolution   Early in their evolution the Chelonia—the tortoises, terrapins, and turtles—were soft-bodied creatures with the normal vertebrate skeleton, and the scaly, protective skin of the reptile.   During their evolution the body scales developed to form a soft shell beneath which the flesh wasted away until the shell fused with the backbone and rib-cage, gradually forming the immensely strong shell that is now common to all land tortoises. Such an evolution meant that the tortoise had to sacrifice speed for strength, and had to modify, for instance, the mechanics of breathing.

The only parts of a tortoise skeleton that can be moved are the limbs, the jaws, and the neck and tail vertebrae. As a tortoise cannot breathe by moving its chest as other reptiles do, it has come to rely on muscular contractions creating changes of pressure within the rigid body cavity, alternately drawing air in and out of the lungs.   Breathing can be very intermittent, with long pauses between one breath and the next.   When a tortoise is suddenly disturbed while holding its breath, air rushes out through the windpipe, making the hissing sound that is a characteristic of so many reptiles.

Survival and Extinction   Certain of the Chelonia have populated Earth from the so-called 'Age of Reptiles'—millions of years before man evolved—right up to the present time.   This extraordinary evolutionary success is always attributed to the shell, but a strong shell is no protection against man.   After surviving 200,000,000 years on this planet, some Chelonia populations have recently become extinct, and others are threatened. Since 1890, the familiar Mediterranean tortoises have been seriously reduced in numbers, just to meet the demands of the pet trade.

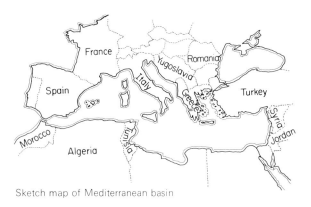

Sketch map of Mediterranean basin

Natural Habitats The spur-thighed (Greek) tortoise is found in many countries of the Mediterranean basin, in addition to Greece: Spain, Romania, Turkey, Iran, Israel, Jordan, Tunisia, Algeria, and Morocco. Before 1978 it was collected in the North African countries, as many as 300,000 being sent here in a single year from Casablanca alone. Latterly, it was exported mainly from Turkey. The spur-tailed (Hermann's) tortoise is absent from Africa, but is found on the European side of the Mediterranean, principally in the South of France, Italy, Yugoslavia, and on many of the islands, including Corsica, Sardinia, and Sicily. Scarcity meant that in recent years it was exported principally from Yugoslavia.

Importation Most informed people were against the continued importation of these tortoises and welcome the EEC ban. Certainly it was a very wasteful trade. Some of the tortoises died on the journey, which they had to endure packed in baskets or crates, without being inspected, watered, fed, or exercised. Many—some estimates say 99% of a consignment—failed to survive one year in Britain. They died during their first winter.

Acclimatization Mediterranean tortoises cannot tolerate temperatures as high as 45°C, or as low as 0°C. If they are exposed to these extremes, death is inevitable.

Suitable box for hibernation

They are accustomed, in the warm temperate countries of the Mediterranean basin to long, hot summers above 21°C, and frequently in their preferred range of 25°–30°C. The winters are short and mild, above 6°C, but even so the tortoises hibernate for a short while since they are inactive at temperatures as low as 6°C. In Britain's cool temperate climate, the summers are normally below 21°C—well out of their preferred range; the winters below 6°C, and frequently below zero. A cool temperate climate, then, is quite unsuitable for Mediterranean tortoises. Those which survive here are much less active than normal, so that they rarely breed here under natural conditions. They also need to hibernate for nearly half of each year to avoid death in our northern winter.

Newly-imported tortoises had only a few months in which to make the transition from a warm temperate climate to a cool temperate, and those imported as late as July could not be expected to do so. It was the first hibernation that was critical. To survive it, a tortoise needed sufficient body fat to sustain it through a long, torpid state. The problem was that body weight was inevitably low after the enforced fast of a long journey. If we were enjoying a good summer they might feed well and put on enough fat to last them through an approaching hibernation of perhaps 5 months' duration, but if temperatures here were low, then they would lapse into torpor and feed inadequately.

Body Temperature Tortoises, together with all other reptiles, are poikilothermic—their body heat varies with that of their surroundings. Imported into a cool temperate climate, tortoises that have been active a few weeks previously in the Mediterranean can immediately lapse into torpor. By keeping them indoors, or in a vivarium kept at temperatures within their preferred range, they could be induced to feed again, but unless newly-imported tortoises received this help in feeding from understanding owners, they are unlikely to have survived their first hibernation here.

It is worth realizing that reptiles do, to some extent, regulate their own body temperature by basking in, or by avoiding the sun. In particular, tortoises need freedom to position themselves—often tilted against a warm wall for maximum effect—to get the benefit of the morning sun for an hour or so to reach a temperature at which they are active enough to feed. Conversely, on hot summer days they must have access to shelter, preferably beneath cool vegetation, to keep down their body temperature. On

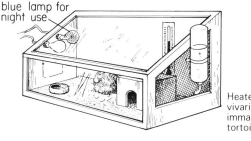

blue lamp for night use

Heated vivarium for immature tortoises

159

really hot days a cool, shallow bath, and a shower from a watering-can will help keep them safe and comfortable.

Hibernation   The spade-like limbs of a tortoise are well shaped for digging, and the claws of some older specimens show noticeable wear.   In nature they burrow into the ground for their hibernation, but this cannot be recommended in captivity.   Hazards such as their burrowing into a bonfire instead of the ground; exposing themselves to frost by not burrowing deep enough in hard ground; being accidentally speared by a garden fork in the spring; of emerging in a neighbour's garden when they do wake— all make it necessary for the owner to supervise the hibernation.   Although it is dependent on the weather, tortoises in this country have usually stopped feeding by November.   It is said to be very unwise to allow a tortoise freedom to move around once it has stopped feeding, for it is merely using up valuable body fat.   The tortoise needs to be put in the middle of a large, stout box, packed with straw, shredded paper, dry autumn leaves, or any similar insulating material, and the box fitted with a strong, but ventilated lid.   It is best kept in a cool place such as an unheated garage, outhouse, or shed where frost cannot penetrate.   Frost is a common killer, but other tortoises are attacked by rats and mice, so every precaution needs to be taken to protect a helpless pet from these dangers.

In recent years, exporters voluntarily agreed to send us only tortoises with a plastron length in excess of 10 cm.

It is possible to save a small, or lightweight tortoise from certain death during the winter by keeping it, feeding normally, in a heated vivarium.   But there is a word of warning.   In laboratories, Mediterranean tortoises deprived of hibernation in this way for several years consecutively show a very high death rate.   It seems that hibernation is an important part of their normal body rhythm, and undue interference is apparently fatal.

Springtime   When they rouse from hibernation in the spring, tortoises need to be cared for indoors for a few weeks to warm up and again reach an active state. Attempt to bathe open the eyes and mouth, and give a warm bath in a shallow dish of clear water.   The shell may be rubbed with olive oil.   As its body temperature

Garden shelter for tortoise

rises above 15°C, a tortoise will begin to feed again. It is most risky to leave a tortoise, once roused from hibernation, to lapse back again. By spring it will have used up practically all its reserves of body fat and, unless it can feed soon, will succumb to death by malnutrition although it seemed to have survived the winter. As the weather improves, the tortoise can spend longer and longer in the garden, but will need to be boxed up at night until the danger of frost has passed.

Drinking Water   After hibernation, tortoises will normally drink before they feed.   Water should be provided, as at other times, in a shallow or sunken dish.   One may see tortoises drinking, bird-like, from a natural puddle in the garden, but they have difficulty in drinking from a bowl.   The tongue cannot protrude from the mouth to lap water, and a small spur on the plastron prevents their lowering the head straight down.   They need to reach forward towards their drinking supply.   If a sunken dish quickly becomes muddy, a saucer may be used instead, stabilized by a few stones pushed under its rim.

Feeding   Of all the reptiles, the slow-moving tortoise, hampered by a heavy shell accounting for about 30% of its body weight, is the least well equipped to take live animal food. When opportunity arises it will take slow-moving invertebrates such as slugs—as country people have rightly said for years—but is predominantly herbivorous. Tortoises have no teeth, but they use their horn-covered jaws to tear food into pieces small enough to swallow. They feed most easily when they can pull against the weight of a whole cabbage or lettuce, or a well-rooted plant, but they do manage to steady a piece of fruit or a single leaf with one fore limb as they eat. Favourite foods include pulses such as peas and beans; greens such as lettuce, cabbage, grass, plantain, vetch, and groundsel; root vegetables such as carrot, turnip or swede—all offered grated if too hard for them to bite; fruits such as apple, pear, banana, plum, strawberry, and tomato—cut in half to give a biting edge; flowers such as rose, pansy, nasturtium, poppy, clover, sedum and dandelion. Individuals will show certain preferences, but most tortoises eat the natural succession of foods as they become available throughout the year beginning, often, with dandelion flower and fallen blossom in the spring. This diet should be supplemented both nutritionally, and in calorific value, with a little cereal food in the form of brown bread, breakfast cereal, Bemax, or Farex; some animal protein in the form of cat or dog food; and calcium in the form of bone meal sprinkled on other food, or cuttlefish 'bone' either crushed, or given whole. As always, a varied diet is likely to be the most well balanced. There is no danger of a tortoise overeating in our climate. Indeed, considerate owners will keep a tortoise in the house during a cold spell, to encourage it to continue feeding enough to survive the approaching winter.

Exercise Ideally, tortoises should be allowed to feed freely in a well-bounded garden, selecting most of their own food, but fenced off (p. 163) from areas poisoned by the use of herbicides or insecticides, and from areas where there are precious plants. It is necessary to make sure, several times a day, that they have not, with their ambitious climbing over a rockery or rough ground, fallen helpless upside down. Tethering cannot be recommended as the string may damage the shell, cut deep into a limb or even strangle a tortoise. The choice facing most tortoise owners with a garden is that of giving their pets a dull but safe life in an enclosure, or of allowing them freedom, with its associated hazards. Those who have no such choice—people who live in flats or town houses—may be able to keep tortoises if they have a balcony or paved area that receives sunlight at some time of day, where the tortoise may have a certain amount of freedom. The place will have to be kept hygienic by frequent and thorough cleaning. In such a situation tortoises will need a shelter (similar to that given to garden tortoises in an enclosure), where they may escape rain, too hot a sun, and the cold of the night.

Breeding It is as well to keep a pair of tortoises, even if the chance of their breeding here is remote. Apart from the advantages of companionship, the presence of the female will stop a male from wandering off to seek a mate when we do have warm summer weather. Adult weights are about 850 g for males; 1600 g for females. The tail, too, is an indication of sex; the male's is longer than the female's, and relatively thicker at the base, where the penis is contained. The male's plastron is concave, fitting over the dome of the female's carapace during mating, for the eggs are fertilized in her body. But the females are also capable of laying infertile eggs, as are birds, although any laid within 3 or even 4 years of her importation could be

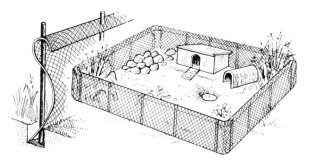

Tortoise enclosure with inset showing detail of fence

Correct
handling
method

fertile as a result of a copulation when she was still in the Mediterranean area. When found, the egg tops should be marked so that they may be kept the right way up, without turning, buried in a tray of dry sand for as long as 20 weeks within the range of 25°C–30°C. In the wild, a clutch of about 6 eggs is laid in a hole dug by the female, but the babies receive no maternal care, so any successfully hatched here are not at a disadvantage in that respect. The shell is soft on hatching, and very gentle handling is important, but if the young are kept in a vivarium within the preferred temperature range; fed a highly nutritious diet of finely shredded food, with plenty of calcium for good shell development; allowed the benefit of direct sunlight; and not allowed to hibernate until the plastron is 10 cm long, then they have a chance of survival.

Conservation It is true that Mediterranean tortoises, once they have lived through one hibernation here, are capable of reaching a great age—60 years or more. They are not likely—except in unusually good summers, or in heated vivaria—to breed here and replenish their kind. This led to constant importation of new stock every year, to meet the demand of the pet trade, with inevitable depletion of the wild populations. For this reason, the 1984 EEC Regulation, banning the importation of Mediterranean tortoises (*T. graeca*, *hermanni* and *marginata*), is welcomed by conservationists throughout Europe. The ban is binding on all the EEC Member States.

Handling Tortoises can easily slip to the ground if picked up from above. It is safer to hold them with the fingers beneath the shell, and the thumbs on top.

Ailments Tortoises are rather subject to COLDS, showing symptoms of laboured breathing, a discharging nose, watering eyes, and loss of appetite. Keep the tortoise within the range 25°C–30°C, bathing the eyes, and keeping the nose clean, and if there is no improvement after a week seek veterinary advice, since there may be a danger of PNEUMONIA developing.

A parasite—the TICK—is not now so common as previously, but any found on a tortoise's limbs should be removed by dabbing with a few drops of liquid paraffin or methylated spirit, applied with a child's paintbrush. Gradually this treatment will cause a tick to loosen its hold until it can be pulled clean away with a pair of tweezers. Tortoises can also suffer from INTERNAL PARASITES in the form of worms. Veterinary treatment is recommended.

TORN CLAWS can be filed smooth; overgrown ones may be worn down naturally if the tortoise is exercised on concrete or some similarly abrasive surface. A veterinary surgeon would use bone forceps to cut seriously overgrown claws; a careful owner can successfully use sharp electrician's wire cutters.

No pet should be destroyed by an unskilled person. This is particularly true of tortoises because they cannot be killed humanely in a chloroform chamber. If necessary, a veterinary surgeon will destroy a tortoise painlessly by injection.

Terrapins

Emys orbicularis and *Chrysemys scripta elegans*

In Britain, land forms of the Chelonia are called tortoises; marine forms are called turtles; freshwater forms are called terrapins. When confusion arises it is partly because in the USA the word 'turtle' is used for all the Chelonia. Our pet-shops normally stock two distinct kinds of terrapin, each needing very different care. One is the European pond terrapin, *Emys orbicularis* (Linné 1758), often called the European pond tortoise; the other,

Shell of red-eared terrapin

sub-tropical species of which the red-eared (or elegant) terrapin, *Chrysemys scripta elegans* (Weid. 1839), often called the red-eared turtle, is most often seen.

Origins Of the two kinds, the European pond terrapin, imported from the Mediterranean area, is the most likely to survive here. It was once indigenous to this country. Modern attempts to reintroduce it have failed, but natural populations are still found as far north as Germany.

Description Predominantly brown in colour, it has the typical terrapin outline of flattened carapace, long tail, and webbed toes. Perhaps the most interesting feature, biologically, is the hinged plastron: two lobes—anterior and posterior—which can be raised to encase the complete terrapin within a safe box. The seal between the carapace and the raised plastron is not as firm as in the true box Chelonia, the *Terrapene*, of which the Carolina box turtle is the archetype, but the European pond terrapin is sometimes classified as a semi-box type.

Breeding The sexes are similar in size, and capable of growing to 30 cm or more. The male's plastron is concave, and his tail longer than the female's (cf. tortoises p. 163). They mate in water, although the eggs are laid on land. Any eggs found should be treated like those of the Mediterranean tortoise (p. 164), but kept in damp sand.

Accommodation Adult European pond terrapins are

166

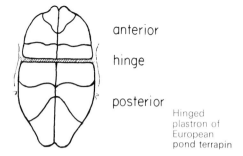

anterior

hinge

posterior

Hinged
plastron of
European
pond terrapin

best kept in an outdoor pond, with an island or shore where they may bask in the sun. Their preferred temperature range is 24°–27°C, but mature specimens tolerate our summers well. In general, terrapins and turtles have shed some of the weight of the shell during evolution, making them more mobile than tortoises. For this reason, a terrapin pool will need a fenced area around it to contain agile terrapins which, although normally considered aquatic, like to spend quite a lot of time on land, taking to the water when disturbed, and to feed, to sleep, to cool down, and to avoid bad weather.

Immature European pond terrapins, just a few centimetres long, should be kept indoors in an aquarium, the recommended temperature range being as high as 27°–30°C.

European pond terrapins must have a pond

European pond terrapin, *Emys orbicularis*, laying eggs

Feeding Although they will take some water plants, and lettuce leaves floated on the water, these terrapins are essentially carnivorous and should not be kept in a pond with fish. They will take small aquatic animals such as insects and tadpoles, and will tear any large prey offered—for instance, herring—into pieces that can be swallowed. Red meat, liver and other offal is also recommended for their diet. Immature specimens will need to be fed *Tubifex*, *Daphnia* and mealworms, together with scraped or very finely chopped liver. Adult terrapins are usually fed one meal a day; immature ones are best fed several small meals.

Cleaning Although whole animal food is essential nutritionally, the uneaten portions do foul the water. This does not matter so much in a large pond, where there are probably enough natural scavengers to clean up after a pair of terrapins, but it is necessary to be able to drain and clean a small pond.

Hibernation In nature, the European pond terrapin hibernates in the mud at the bottom of a natural pond or lake, but in captivity it should be allowed to hibernate like a tortoise (pp. 159–60), except that the box should contain damp earth, loosened by being mixed with autumn leaves.

Ailments The most likely ailment is SOFT-SHELL, due to

a deficiency of calcium and vitamin D. Avoid it by feeding fish liver oils on food, particularly in winter; allow the terrapins access to natural sunlight; and include whole fish, and also cuttlefish 'bone' in the diet.

RED-EARED or ELEGANT TERRAPIN

Origins   Young specimens of this beautiful creature are imported from the Gulf states of the USA (New Orleans Jan. 12°C; July 28°C, where they live naturally in ponds, ditches, swamps, and lake margins but are 'farmed' and exported by the million for the pet trade.

Description   When warm these terrapins are extremely active, scuttling about on land, and swimming rapidly using only the back legs. The young are bright green with thin yellow lines on the head and neck, and a conspicuous red spot behind each eye. With age, the males in particular lose their colour, but because so many youngsters die here, mature specimens are not often seen.

Breeding   The red-eared terrapins mature at the relatively early age of 3 years. During courtship, the little male swims backwards before the larger female, stroking her face with the exaggerated claws of his fore limbs, and displaying more finesse than the Mediterranean tortoises and terrapins with their shell-butting and leg-biting.

Accommodation   In this country red-eared terrapins must be kept indoors in a vivarium heated to their preferred temperature range of 25°C–30°C, although the temperature may be allowed to drop a few degrees at night as it would in the natural habitat. Whenever possible the vivarium must be placed where the sunlight can fall directly onto the terrapins. If the sunlight first passes through glass, the ultra-violet light is filtered out and the terrapins are liable to suffer from soft-shell.

Hatchlings need only shallow water, but adults require enough for fast swimming. It is not usually possible to decorate the vivarium as one would a fish tank, for every day the water will need changing and replacing with clean water at a similar temperature. This is an unavoidable chore for most owners, although there are two possible ways of avoiding it. Sometimes it is possible to keep a

Underside of red-eared terrapin, *Chrysemys scripta elegans*

separate small pool, such as pet-shops sell for hatchlings, just as a feeding pool. The objection is that small terrapins may return to a meal several times a day, and could go hungry if restricted to particular feeding times. However, careful observation of chosen feeding times may reveal a pattern so regular that one would feel quite confident in using a separate feeding pool.

Another suggested way of avoiding disturbance of the entire vivarium each day is by having the water in a separate container that can be drained, or removed for cleaning, quite independent of the 'shore' area. In this, as in all difficult matters of pet-care, the skilful owner will display his own flair.

Feeding Although mostly carnivorous, the red-eared terrapin will also take some green plant food. Small hatchlings should be given finely shredded food—scraped red meat, and so on—but it is a mistake to feed them only animal flesh. For a nutritious diet they need the entrails and crushed bones as well. For this reason, liver and

Heated vivarium suitable for red-eared terrapins. Terrapins breathe through lungs and must be able to reach an 'island' or 'shore' area above the water level.

other offal, and also crushed cuttlefish 'bone' are desirable in the diet. Supplements such as fish liver oils may be given wrapped in tiny pieces of meat that will be swallowed before being tasted. This is easy if the terrapin will take food from the hand, but it takes all one's ingenuity to feed adequately a terrapin that is less tame. For this reason the hatchlings are not suitable for any but the most devoted pet-keepers.

Ailments REFUSING FOOD may be an indication that the temperature is too low for this poikilothermic animal (see tortoise section p. 159) to feed adequately, and unless corrected will result in death from malnutrition. It may also be a symptom of SOFT-SHELL, which is a prevalent disease among captive Chelonia denied access to sunlight. It is similar to rickets in man, and prevents proper shell formation. The dietary fault is insufficient calcium and vitamin D, and a terrapin that refuses all food for several weeks is probably suffering from this disease in its terminal stage, and should be taken to a veterinary surgeon who may recommend destruction. A high proportion of terrapins also suffer from EYE COMPLAINTS and when these are not caused by an infection, or poor, diet, then it is suggested they may be due to hard water.

FISH

The Goldfish   *Carassius auratus*

The Common goldfish is remarkably hardy, and in spite of an alarmingly high mortality rate numbers among the most commonly kept pet animals in the country.   It has been kept as an ornamental fish for centuries, and graced the garden pools of kings and emperors for a thousand years before suffering its present ignominy of being given away in plastic bags at summer fêtes.

Description   The Common goldfish, *Carassius auratus*, Linné, renowned for its spectacular red-gold colour, can grow to about 40 cm in length, and live for 25 years or more.   A typical fish, it has an arched, streamlined body with no neck, breathes dissolved oxygen by means of gills, and controls its movement through water by means of fins.

There are two sets of paired fins—the pectoral and pelvic—which act as brakes when raised.   When turning, a goldfish can be seen to raise one pectoral fin to exert a drag on that side only.   An anal and dorsal fin act as keels, affecting the trim of the fish, and a caudal fin is used in swimming, although the main thrust is generated by body movements.

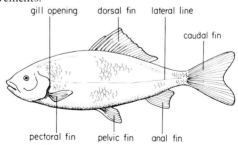

Common goldfish

172

The body is protected by rows of overlapping scales that never increase in number, but increase in size as the fish grows. The circuli, or rings of growth, are well-spaced in summer when growth is rapid, but are laid down close together in winter and at other times of poor growth. The tightly-bunched circuli of winter tend to show up as distinct bands, or annuli, and it is these that indicate the years of a fish's life to a trained scale-reader.

The lateral line is a row of specialized scales running from head to tail each side of the fish. It is a major sense-organ, registering even the most sensitive vibrations in water, the slightest changes in water pressure and velocity, and much more than we understand besides. It allows the fish to dart unerringly into small rock-crevices, to execute high-speed turns in a shoal without collision, and to swim fast in the confines of a new aquarium without ever touching the glass sides.

The eyes of the Common goldfish protrude slightly, effectively giving all-round vision of prey and predators to an animal that cannot turn its head. They cannot be adjusted in any way—not even closed—since there are no eyelids, and are entirely unprotected.

Internally, there is a highly flexible backbone with spines radiating from both sides, corresponding to the ribs of other vertebrates, but not forming a corresponding rib-cage to contain vital organs. The major organs—heart, kidneys, liver, for instance—compare with those of other vertebrates. In addition there is a buoyancy chamber—an air-filled swim-bladder.

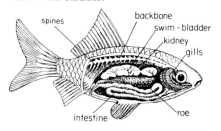

A fish with the body wall and pectoral fin removed to show the principal internal organs

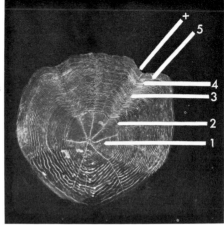

Goldfish scale. The numbers indicate years of growth

The most obvious difference between the internal organs of fish and other vertebrates is the presence of gills in place of lungs. Fish breathe by circulating water through the gills. The exchange of gases takes place as the water flows over the gill filaments; dissolved oxygen passes from the water into the bloodstream and carbon dioxide from the bloodstream into the water. The water is taken in at the mouth and expelled by way of the gill slits. The gill covers (opercula) can be seen to rise and fall with the flow. The nostrils are organs of smell and are not used at all in breathing.

Varieties From the Common goldfish more than 100 varieties have been bred, collectively called fancy goldfish. These display extraordinary divergence of form, typical of long domestication, but in general tend to be smaller, less hardy, and to live shorter lives than the Common goldfish. The longest lived of them (Comet, Shubunkin, Fantail) have a life span of 14 years. 'Visibly-scaled' varieties have scales as distinct as those of the Common goldfish; 'calico' varieties have such inconspicuous scales that they appear scaleless.

The Comet most resembles the Common goldfish, but has a more streamlined outline, and an exaggerated caudal

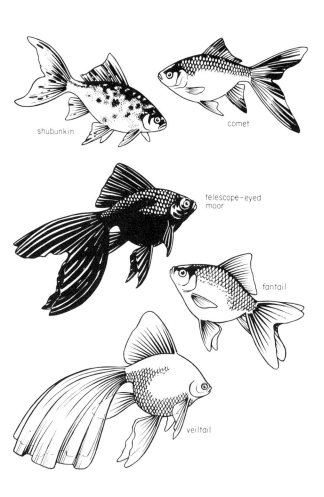

shubunkin

comet

telescope-eyed moor

fantail

veiltail

Fancy goldfish

fin that equals the length of the body. There are two varieties of Shubunkin, both calico: the Bristol, with the outline of the Comet, and the hardier London variety, with the outline of the Common goldfish. Good specimens of each are prized for their vivid blue bodies, flecked with black, and mottled with patches of red, brown, yellow, and violet.

The Fantail has an egg-shaped body with paired anal fins, and a distinctive forked tail that is doubled for most of its length. Similarly, the round-bodied Veiltail has paired anal fins, and a double tail that hangs in folds but is not forked. Calico varieties of Fantail and, more rarely, of Veiltail are sometimes seen, both renowned for having the multicolouring of the Shubunkin in place of the normal gold colouring.

A jet black fancy goldfish is the Moor, which closely resembles the Veiltail in outline and fins, but has protruding, telescopic eyes. The Oranda too has the body and fins of the Veiltail, but its distinguishing feature is a mane-like growth on the head. As might be expected from the name, this growth is much more pronounced on the Lionhead, which again has the body of the Veiltail. This fish also lacks a dorsal fin, although its other fins compare with those of the Fantail. Both Orandas and Lionheads retain the normal goldfish colouring.

Domestication The Common goldfish originated in East Asia and was greenish-bronze in colour. It was able to survive in water poor in oxygen, and to breed in small ponds. It was therefore the obvious choice of the Chinese who, about 1000 years ago, were seeking a fish that could be farmed for food production.

With selective breeding, size increased dramatically, but as the first gold mutations appeared, the economic value of the fish changed from that of a food animal to that of an ornamental animal. Breeders began to concentrate on improving the goldfish, and breeding for food was abandoned. Goldfish breeding continued in China, and also in Japan, for 700 years, and began in Europe in 1691 when the first goldfish were brought here by sailing ship. The first to breed in this country were imported in 1728.

Common goldfish

Accommodation: Ponds Ponds are the only acceptable habitat for the larger goldfish, providing these are hardy enough to tolerate winter temperatures. The Common goldfish, London Shubunkin, Comet, and visibly scaled Fantail may be kept in a garden pond all year round in Great Britain; the Bristol Shubunkin, calico Fantail, Veiltail, Oranda, and Lionhead are safe only during the summer.

Other compatible species of coldwater fish include Golden Rudd, Tench, and Golden Orfe. The Golden Orfe is particularly valuable in a pond for it is a surface feeder, and so active that it is seen constantly. Its colouring is pale gold, usually with black markings. The Golden Rudd is a bottom-feeder, but also active enough for a flash of its distinctive red fins to be seen from time to time. The other useful bottom-feeders are the Tench. These serve the utilitarian purpose of scavenging food wasted by the surface feeders, and so keeping the pond clean. They keep to the bottom always, and are seldom seen, although there is a golden variety that shows up a little better than the green.

Temperature Fish are poikilothermic (cold blooded), entirely dependent on their surroundings for their body heat, and quite unable to tolerate sudden changes of temperature. This is no problem in nature, because in large bodies of water there is much less variation in temperature than on land, but in captivity it can prove fatal. Small garden ponds are subject to sudden changes in temperature, and shallow ones can freeze solid. In order to survive outside all winter, fish must be protected by a sufficient depth of water.

Depth If a garden pond is constructed with one deep spot of about 1 m, the temperature there is unlikely to drop below 4°C even when there is a thick layer of ice at the surface. Goldfish, and the other recommended species, are able to survive under the ice in a torpid state, the only problem being shortage of oxygen. Smashing the ice can do more harm than good, by setting up vibrations that shock the fish; gentler methods of opening up small air holes can be recommended. One such is to float balls on the pond at night, and remove them during the day when temperatures rise above freezing.

It is not necessary for the whole pond to be 1 m deep. In fact, a range of depths allows more versatility in planting, and some shallow water is essential if the fish are to breed.

Pre-formed ponds The simplest garden pond is made by using a pre-formed plastic or fibreglass shape which is sunk into a hole excavated to fit. These ponds last well, but fibreglass is expensive. The disadvantage is that, to keep down costs, owners are often tempted to buy too small a size that gives insufficient shelter in winter, and inhibits the growth of the fish. The small sizes are best used only as temporary, summer ponds.

Polythene-lined ponds A much cheaper pond can be made by excavating the required size and shape, and lining it with two thicknesses of heavy duty black polythene sheeting, or a manufactured pool liner. The excavated hollow will have to be covered with sand or sifted soil to prevent sharp stones from puncturing the polythene when the water is added. Before introducing the water stretch the polythene across the hollow, following the main curves of the excavation, and anchor it around the sides with heavy stones. There needs to be a 30 cm surplus around the perimeter at this stage. As water is slowly introduced by means of a hosepipe, the

Goldfish in lily pond

polythene will stretch slightly to make a good fit. This is important, for loose folds in the polythene can trap fish. Finally the surplus polythene is cut off, and the edges of the pond planted with turfs or covered with paving stones. Such a pond will last well with careful use, but will need re-lining after 7–10 years as the polythene deteriorates.

Maintenance The amount of maintenance required is very much dependent on the pond's location. Those overshadowed by trees will need netting in the autumn, or the leaves raked out annually—a hazardous chore in a plastic pond. Decaying leaves deplete the oxygen level during decomposition.

The development of algae is hindered if the water is shaded, for instance by water-lilies, or floating plants such as *Azolla.* In full sunshine algae will grow profusely. 'Blanket weed' is particularly difficult to control. If hand-weeding fails, the only treatment is to remove all fish and plants temporarily while the algae is killed off with a strong dose of potassium permanganate. A dark red solution will kill blanket weed in 3 days, but would also prove fatal to fish and plants. Afterwards the pond must be carefully drained, rinsed, and re-filled. Take care not to reintroduce the algae with the plants.

Accommodation: Aquaria   Recommended species of goldfish for aquaria are young specimens of Common goldfish, Comet, London Shubunkin, and visibly-scaled Fantail, all measuring less than 12 cm; the more delicate Bristol Shubunkin, calico Fantail, Veiltail, Oranda, and Lionhead during the winter; and the Moor which should always be housed in an aquarium.   The Moor is too short-sighted to feed well in a pond, and its protruding eyes are liable to become damaged.

Aquarium tanks   Three types of tank are generally available: clear, moulded plastic; stainless steel framed; and angle iron framed, preferably nylon-coated.   A cheaper alternative can sometimes be found—a large plastic washing bowl or baby's bath—but under no circumstances is a goldfish bowl satisfactory.   If there is no hood a sheet of glass, raised at the corners on four balls of plasticine, will allow air to circulate over the surface yet keep out dust, and prevent fish from jumping out.   The number of fish which can be housed adequately depends on the surface area.   For small fish under 8 cm long allow 24 square cm of surface water for every 1 cm of fish (i.e. body length, excluding caudal fin).   Larger fish need rather more room.

Aeration   The fish capacity of a tank is increased by as much as 40% by continuous aeration.   An air-pump is not necessary for a coldwater aquarium, but once fish have become accustomed to aerated water it should be continued.   The advantage is that apart from bubbling air directly into the water, the stream of bubbles moves lower levels of water towards the surface where carbon dioxide can be given off, and oxygen dissolved.   Exactly the same effect is achieved in a pond by the use of a fountain pump.

Position   The most successful aquaria are placed in a north light—although not on a window sill where temperatures may fluctuate too much—so that neither fish nor water is subjected to strong sunlight which is too bright for a fish's eyes, and encourages too much algal growth. If shade is needed the back, and perhaps the sides of the tank can be masked with dark, thick paper, or a bought frieze depicting an underwater scene.

Furnishing   Furnish the tank with a 5 cm layer of well-washed aquarium gravel, and use rounded rocks to make

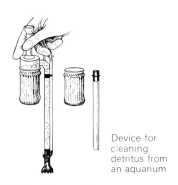

Device for
cleaning
detritus from
an aquarium

hides for the fish.   Sharp-edged rocks, shells, sea urchin
tests, and anything else that might cause damage to the
scales or fins of the fish should be rejected.   Water plants
can be weighted with stones or lead strip, and anchored in
the gravel.   Before adding tap water—very slowly—it is
worth covering the plants temporarily with a sheet of
plastic or newspaper to minimize disturbance of the
gravel.   Tap water will usually warm up to room
temperature and lose its chlorine in 24 hours, but allow
plants to settle for a week before first introducing the fish.

Heating and lighting   No artificial heating or lighting is
needed for the fish.   Room temperature in this country
can be expected to remain within the goldfish's preferred
temperature range of 10°C–21°C.   The plants, however,
thrive best on artificial top lighting for 8–10 hours a
day.   Owners are warned that fish can suffer shock if
suddenly plunged from dark into light, or *vice versa*.   A
more gradual transition can be achieved by lighting the
room before lighting the aquarium; and darkening the
aquarium before the room.

Maintenance   Frequent maintenance tasks are scrap-
ing algae from the glass, using an aquarium vacuum
cleaner to remove uneaten food and other waste (detritus)
from the bottom of the tank, and polishing up the hood
and the outside of the tank.   Less frequently, plants may
need to be trimmed and tidied, and some water siphoned
off from the bottom of the tank and replaced with tap
water that has stood for 24 hours.

Removing algae with scraper

Water Plants Plants are grown in ponds and aquaria mainly for decoration, for shade, and for their ability to absorb carbon dioxide and many of the waste products of decomposition such as ammonia and nitrates. They are less valuable as oxygenators than is generally supposed. Oxygen is only given off during their feeding process of photosynthesis. Throughout the 24 hours, oxygen is used by the plants for respiration. Suitable submerged plants for ponds and coldwater aquaria are:

Ceratophyllum demersum	hornwort
Elodea canadensis	Canadian pondweed
Fontinalis antipyretica	willow moss
Myriophyllum spicatum	water milfoil
Potamogeton crispus	curly pondweed
Sagittaria sagittifolia	arrowhead (pond)
Vallisneria spiralis	eel grass (aquarium)
Ranunculus aquatilis	water crowfoot (pond)

In addition, a specialist nursery will be able to supply floating plants, water-lilies in their special planting baskets, and a large range of sub-aquatic plants for the pond margins.

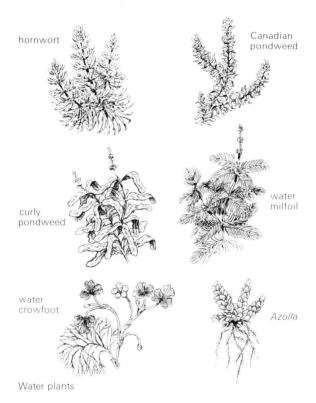

hornwort

Canadian pondweed

curly pondweed

water milfoil

water crowfoot

Azolla

Water plants

Feeding Coldwater fish feed best in a temperature of 16°C–18°C. Outside their preferred temperature range of 10°C–21°C, they will probably cease to feed, either because of their inactivity in cold water, or inactivity in water low in oxygen (as water warms it holds less oxygen in solution). It follows that the amount of food given must vary with the temperature. In summer perhaps 2 meals a day; in winter 1 meal every other day may suffice. The use of a feeding ring confines the food to one spot, and

Water-lily and planting basket

makes cleaning easier. The correct amount of food to give at a meal is the amount the fish can finish in 10 minutes. Overfeeding is dangerous. Uneaten particles foul the water and, as they decay, deplete the oxygen level. In good condition fish can survive 2 or 3 weeks without food—and longer if water plants are available—but they very soon die in foul water.

By nature goldfish are omnivorous, feeding on aquatic insects, small fish, and green plants. This natural diet is still available to them in a well-stocked pond, supplemented as need be. Aquarium fish are more often fed a proprietary dried fish food high in protein, with some fresh or frozen food once or twice a week for variety. Dietary supplements in the form of fish tonics are also available, and if water plants are not available, a little chopped greenstuff, such as lettuce, must be fed.

The simplest fresh food is a small piece of raw meat—well-washed to remove excess blood—suspended in the water. Other suitable fresh foods include the water flea, *Daphnia*, the freshwater shrimp, *Gammarus*, and the freshwater louse, *Asellus*. These, together with *Tubifex* worms, are available as frozen diets.

Breeding The spawning season lasts for the whole summer beginning in April or May. From the age of 1 year, healthy goldfish are capable of spawning every month of the season. In breeding condition the sexes can be distinguished: the male has white tubercles on the opercula and pectoral fins; the female is swollen by the ovary (i.e. hard roe) containing thousands of eggs. One male is able to fertilize the eggs of many females.

The unmistakable breeding behaviour, in which the males chase the females into the warmest, shallowest part of the pond, is accompanied by much noisy splashing and may last all day. The chase induces the females to shed their eggs, which are immediately fertilized by the sperm or milt of the males and adhere to water plants.

In nature many eggs and fry (hatchlings) are eaten by water snails and adult fish, and they have a much better chance of survival if a plant to which they are clinging is removed to a clean tank with sides shaded from the sun. At 21°C the fry will hatch in 4 days; at 10°C hatching will take a more hazardous 14 days.

Once the fry swim freely their natural food is microscopic animal life called infusoria. In a tank they can be fed a proprietary food specially prepared for fry. At the age of 1 month the fry can make the transition to the adult diet; at 3 months the hardy goldfish (Common, Comet, Fantail) can be returned to the pond.

Breeding in an aquarium differs only in that normally just one pair of fish is used. The eggs are best transferred to a separate tank, but alternatively some can be protected from the parents by fixing a glass partition in the aquarium after spawning.

Handling It is vital to protect the scales and fins from damage. In a tank, use a jug or net to catch goldfish against the glass; in a pond, use one net to drive them into another.

When transferring fish a thermometer should be used to check that the water temperatures are similar, as a sudden variation of only 1°C can prove fatal. A newly-bought goldfish is best floated in the aquarium in its plastic bag until the temperatures correspond. A goldfish won unexpectedly is best released into a basin and stood in a cool, quiet place, with no food for 24 hours. Its meagre supply of water can be gradually supplemented with small volumes of tap water aerated and warmed by vigorous shaking. Meanwhile, a bucket of tap water is given time to stand for 24 hours. The goldfish can then be accommodated temporarily in the bucket while a pond is dug or an aquarium set up.

Ailments GAPING at the surface indicates serious lack of oxygen—perhaps from overcrowding, hot weather, or detritus—and must be dealt with immediately. A pond can be aerated by playing a hosepipe on the surface.

When scales are accidentally rubbed off, regenerative ones seldom have time to grow before the white fungus disease SAPROLEGNIA gains entry. The disease causes death when the fungus reaches the opercula and impedes respiration. Isolate affected fish, as the disease is highly contagious. Early treatment is the traditional salt (sodium chloride) bath; the fish is transferred to a 3% saline solution for 15–20 minutes a day. Alternatively, the fish can be transferred to a bowl containing 5 drops of 2% mercurochrome* solution per 5 litres of water for 3 hours a day. In advanced cases the disease will respond only to drugs available on veterinary prescription.

Another common, highly contagious, and often fatal disease is WHITE SPOT or *Icthyopthiriasis*, caused by a protozoan parasite. Isolate visibly scaled varieties in a tank without plants, and attempt to kill off the protozoan by colouring the water blue-black with methylene blue for 8–10 days. Alternatively, the mercurochrome treatment is effective, and is recommended for calico varieties.

Many fish ailments result from poor feeding, overcrowding, foul water, and temperature fluctuations. Trailing faeces indicate CONSTIPATION, which usually results from feeding dried food exclusively. Treatment is to include green and fresh food in the diet, to give a fish tonic, and in cold weather to increase the temperature to 16°–18°C. Similarly TAIL and FIN ROT is cured only when living conditions are improved. The salt bath treatment is usually helpful.

LOSS OF BALANCE may indicate a defect of the swim-bladder for which there is no cure, but constipation, indigestion, and temperature fluctuation all cause the same symptom. Balance can often be recovered by treating the constipation, moving the fish into shallow water for several days, or by keeping the temperature static.

DISCOLORATION and fading may be an early symptom of deteriorating health, or may be caused by chlorine in the water. Fry do not attain adult coloration until 8–12 months, and some fry never develop a good colour.

---

* Mercurochrome is a strong poison, but available in solution from a chemist. An overdose will kill.

PLANTS POISONOUS TO ANIMALS

Aconitum anglicum	monk's hood
Aesculus hippocastanum	horse-chestnut
Aethusa cynapium	fool's parsley
Anagalis arvensis	scarlet pimpernel
Anemone nemorosa	wood anemone
Arum maculatum	lords-and-ladies
Atropa belladonna	deadly nightshade
Buxus sempervirens	box
Chelidonium majus	greater celandine
Clematis vitalba	travellers'-joy
Conium maculatum	hemlock
Convallaria majalis	lily-of-the-valley
Convolvulus spp.	bindweeds
Datura stramonium	thorn-apple
Digitalis purpurea	foxglove
Delphinium spp.	larkspur
Dicentra spp.	bleeding heart
Equisetum spp.	horsetails
Euphorbia spp.	spurge
Hyoscyamus niger	henbane
Hypericum spp.	St John's-worts
Iris spp.	flags
Laburnum spp.	laburnum
Leucothoe spp.	laurels
Ligustrum vulgare	privet
Mercurialis perennis	dog's mercury
Oxalis spp.	sorrels
Papaver spp.	poppies
Prunus spp.	wild cherries
Quercus spp.	oaks
Ranunculus spp.	buttercups
Ranunculus ficaria	lesser celandine
Rheum rhaponticum	rhubarb
Rumex spp.	docks
Sambucus nigra	elder
Senecio spp.	ragworts
Sium latifolium	water parsnip
Solanum spp.	nightshades
Taxus spp.	yew

INDEX

189